John Hamblin Smith

Elementary Hydrostatics

John Hamblin Smith

Elementary Hydrostatics

ISBN/EAN: 9783742801722

Manufactured in Europe, USA, Canada, Australia, Japa

Cover: Foto ©berggeist007 / pixelio.de

Manufactured and distributed by brebook publishing software
(www.brebook.com)

John Hamblin Smith

Elementary Hydrostatics

RIVINGTONS

London ... ...   *Waterloo Place*

Oxford ... ...   *High Street*

Cambridge ... ...   *Trinity Street*

ELEMENTARY HYDROSTATICS

BY

J. HAMBLIN SMITH, M.A

GONVILLE AND CAIUS COLLEGE, AND LATE LECTURER AT ST PETER'S COLLEGE
CAMBRIDGE

NEW EDITION, REVISED

RIVINGTONS

London, Oxford, and Cambridge

1872

PREFACE

THE ELEMENTS OF HYDROSTATICS seem capable of being presented in a simpler form than that in which they appear in all the works on the subject with which I am acquainted. I have therefore attempted to give a simple explanation of the Mathematical Theory of Hydrostatics and the practical application of it.

Prior to the publication of this work some copies were privately circulated with a view to obtain opinions from Teachers of experience as to the sufficiency and accuracy of the information contained in it. A few suggestions received in consequence of this arrangement will be found in the Notes at the end of the volume.

I am indebted to several friends for the collection of Miscellaneous Examples given in Chapter VIII. In conclusion I have to express my thanks for the favour with which my attempts to simplify the course of Elementary Mathematics have been received by College Tutors and Masters in Schools.

J. HAMBLIN SMITH.

Cambridge, 1870.

CONTENTS

HYDROSTATICS.

CHAPTER I.

On Fluid Pressure.

1. HYDROSTATICS was originally, as the name imports, the science which treated of the Equilibrium of Fluids, or of bodies in equilibrium under the action of forces some of which are produced by the action of fluids. It is now extended so as to include many other theorems relating to the properties of fluids.

2. A fluid is a substance whose parts yield to any force impressed on it, and by yielding are easily moved among themselves.

3. This definition separates fluids from *rigid* bodies, in which the particles cannot be moved among each other by any force, however great, but it does not separate fluids from *powders*, such as flour, in which we have a collection of particles which can be moved among themselves by the application of a slight force.

4. A fluid differs from a powder in this way: the particles composing a powder do not move among themselves without friction, whereas the particles that make up a fluid move one over another without any friction.

For example, if you empty a mug of flour on a table the friction between the particles will soon bring the flour to rest in more or less of a heap: whereas if you empty a mug of water the particles, moving without friction, run in all directions, and the whole body of water is spread out into a very thin sheet.

5. To distinguish fluids from powders we must therefore make an addition to Art. 2, and we give the following as a complete definition of a fluid.

DEF. *A fluid is a substance whose parts yield to any force impressed on it, and by yielding are easily moved among themselves without friction, and also act without friction on any surface with which they are in contact.*

This definition includes not only the bodies to which in ordinary conversation we apply the terms "fluid" and "liquid," such as water, oil, and mercury, but also such bodies as air, gas and steam.

6. Fluids may be conveniently divided into two classes, *liquid* and *gaseous*. By the term *liquid* we understand an incompressible and inelastic fluid. In reality all fluids with which we are acquainted are compressible, that is, a given volume of fluid can by pressure be reduced in volume. Still so great a force is required to compress to any appreciable extent such fluids as water and mercury, that we may regard them as incompressible in treating of the elements of the subject.

7. The inelastic fluids with which we are practically acquainted approach more or less to a state of perfect fluidity, but in all there is a tendency, greater or less, of adjacent particles to cohere with each other. This tendency is stronger in such fluids as oil, varnish and melted glass, than in such as water and mercury. Hence the former are called *imperfect* or *viscous* fluids.

8. The air which we breathe and gases are compressible fluids, and are endowed with a perfect elasticity, so that they can change their shape and volume by compression, and when the compression ceases they can return to their former shape and volume.

9. Vapours, as steam, are elastic fluids, but with this peculiarity: at a given temperature in a given space only a certain quantity of vapour can be contained, and if the space or the temperature be then diminished, a portion of the vapour becomes liquid, or even in some cases a solid.

10. Before proceeding further with our subject we must explain the meaning of some technical terms which we shall have to employ frequently.

11. A Piston is a short cylinder of wood or metal, which fits exactly the cavity of another cylinder, and works up and down alternately.

12. A Valve is a closed lid affixed to the end of a tube or hole in a piston, opening into or out of a vessel, by means of a hinge or other sort of moveable joint, in such a manner that it can be opened only in one direction.

13. A Prism is a solid figure, the ends of which are parallel equal and similar plane figures, and the sides which connect the ends are parallelograms.

The figure represents a rectangular prism, in which each of the lines bounding the surfaces of the prism is at right angles to each of the four lines which it meets.

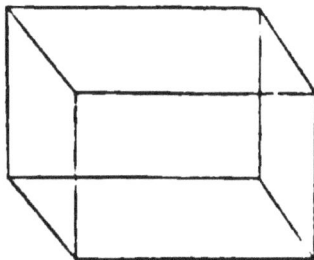

14. We shall often have to use the expression *Horizontal Section* of a tube or hollow cylinder, and we may explain the meaning of the expression by the following example:

Suppose a gun-barrel to be placed in a vertical position: suppose a wad to be part of the way down the barrel with its upper surface exactly parallel to the top of the barrel: then suppose the barrel to be cut away so as just to leave the upper surface of the wad exposed: the area of this surface of the wad is called the horizontal section of the barrel.

15. The mathematical theory of Hydrostatics is founded on two laws, which we shall now explain.

16. LAW I. *The force exerted by a fluid on any surface, with which it is in contact, is perpendicular to that surface.*

17. This law is merely a repetition of the definition of a fluid given in Art. 5, and we can best explain its meaning and application by an example.

If AB be a cylinder immersed in a fluid the pressures of the fluid on the *curved* surface are all perpendicular to the

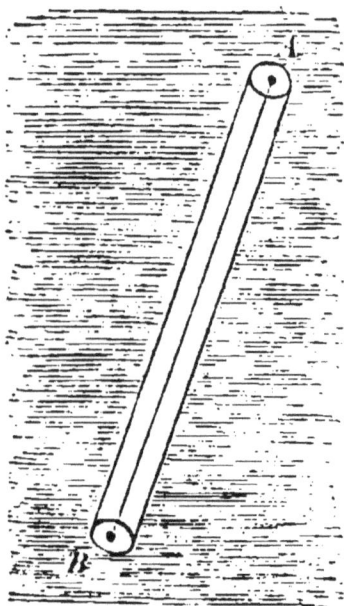

axis of the cylinder, and the pressures of the fluid on the *flat* ends are all parallel to the axis.

Now it is a law of Statics that a force has no tendency to produce motion in a direction perpendicular to its own direction.

Hence the pressures on the curved surface have no tendency to produce motion in the direction of the axis, and the pressures on the flat ends have no tendency to produce motion in a direction perpendicular to the axis.

18. LAW II. *Any pressure communicated to the surface of a fluid is equally transmitted through the whole fluid in every direction.*

19. A characteristic property of fluids which distinguishes them from solid bodies is this faculty which they possess of transmitting equally in all directions the pressures applied to their surfaces.

It is of great importance to form a correct notion of the principle of "the equal transmission of pressure," a principle which is applicable to all fluids, inasmuch as it depends upon a property which is *essential* to all fluids and is not an *accidental* property, as weight, colour, and others.

20. Suppose then we take a vessel *ABCD*, in the form of a hollow rectangular prism, and place it on a horizontal table.

Place a block of wood, cut to fit the vessel, so that it rests on the base *BC* and reaches up to the level *EF*.

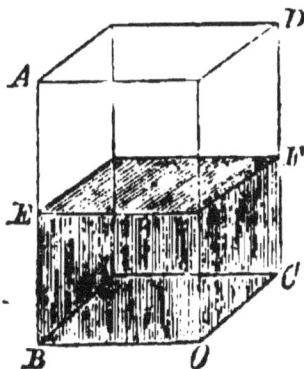

Then if we place a weight *P* on the top of the block an additional pressure *P* will be imposed on the base of the prism.

Now suppose the block to be removed and the vessel filled with an incompressible fluid up to the level of *EF*.

Suppose a piston exactly fitting the vessel to be inserted and a pressure *P* applied by means of it to the surface of the fluid at *EF*.

In this case the pressure *P* is transmitted by means of the fluid not only to the base *BC*, but also *to the sides* of the vessel, and if we take a unit of area, as a square inch, in the side *FO*, and a unit of area in the base *BC*, the same additional pressure will be conveyed to each.

21. *That fluids transmit pressure equally in all directions may be shewn experimentally in the following manner:*

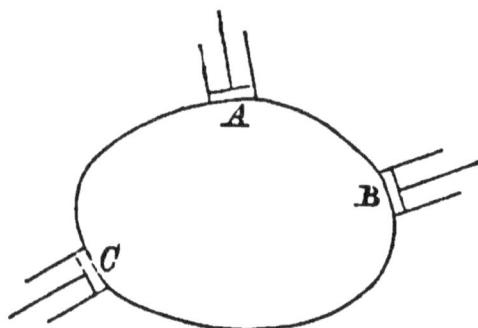

ABC is a vessel of any shape filled with fluid.

Make openings of equal area at A, B, C.

Close the openings by pistons, kept at rest by such a force as may be required in each case. Then it will be found that if any *additional* force P be applied to the piston at A, the same force P must be applied to each of the pistons at B and C to prevent them from being thrust out.

If the area of the base of one of the pistons, as B, be larger than the area of the base of the piston A, it is found that the pressure which must be applied to B to keep it at rest bears the same relation to the pressure applied to A that the area of the base of B bears to the area of the base of A.

22. From the preceding article it is clear that if a body of fluid, supposed to be without weight, be confined in a closed vessel, the pressure communicated to the fluid by any area in any part of the vessel will be transmitted equally to every equal area in any other part of the vessel.

It is owing to this fact that the use of a Safety Valve can be depended on.

Thus, if the vessel A be full of steam and the pressure of the steam be required to be kept down to 200 lbs. on the square inch, if a valve B, whose area is a square inch, be placed at any part of the vessel, and be so loaded that it will require a force of 200 lbs. to raise it, then if the steam acquire an increase of pressure above 200 lbs. on the square inch, the valve will open, and will remain open till the pressure of the steam is just equal to 200 lbs. on the square inch.

23. *Any force, however small, may by the transmission of its pressure through a fluid, be made to support any weight, however large.*

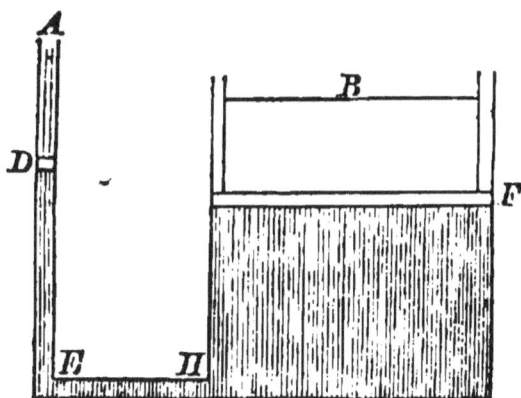

Suppose DE and FH to be two vertical cylinders, connected by a pipe EH, and suppose FH to have a horizontal section much larger than the horizontal section of DE: for instance, let the area of a horizontal section of FH be 400 square inches, and the area of a horizontal section of DE be 1 square inch.

Now if water be poured into the cylinders, and pistons A and B be applied to the surface at D and F, whatever force we apply to A will be transmitted to *each portion* of the base of the piston B which is equal in area to the base of the piston A.

Hence a pressure of 1 lb. applied to the piston A will produce a pressure of 400 lbs. on the base of the piston B, and will therefore support a weight of 400 lbs. placed on the piston B.

This effect of pressure by the medium of a fluid is often called The Hydrostatic Paradox.

Examples.—I.

(1) In the experiment described in Art. 23, if the horizontal section of the small cylinder be $1\frac{1}{8}$ square inches, and that of the larger cylinder 64 inches, find the weight supported under a pressure of 1 ton exerted on the piston of the small cylinder.

(2) If the horizontal section of the small cylinder be $1\frac{1}{2}$ square inches, and that of the large cylinder 240 inches, find the weight supported by a pressure of 3 cwt. applied to the piston of the small cylinder.

(3) If the pistons are circular, the diameters being $1\frac{1}{2}$ inch and 50 inches, find the weight supported by a pressure of 15 lbs. applied to the smaller piston. (N.B. The areas of circles are as the squares of their diameters.)

(4) A closed vessel full of fluid, with its upper surface horizontal, has a weak part in its upper surface not capable of bearing a pressure of more than $4\frac{1}{2}$ pounds on the square foot. If a piston, the area of which is 2 square inches, be fitted into an aperture in the upper surface, what pressure applied to it will burst the vessel?

(5) A closed vessel full of fluid, with its upper surface horizontal, has a weak part in its upper surface not capable of bearing a pressure of more than 9 lbs. upon the square foot. If a piston, the area of which is one square inch, be fitted into an aperture in the upper surface, what pressure applied to it will burst the vessel?

(6) If the horizontal section of the small cylinder be $1\frac{1}{8}$ square inches, and the diameter of the large piston 20 inches, find the lifting power of the machine under a pressure of 1 ton exerted on the piston of the small tube. (N.B. The area of a circle is $\frac{22}{7}$ times the square of the radius nearly.)

24. The pressure *at* any point in any direction in a fluid is a conventional expression used to denote the pressure on a unit of area imagined as containing the point, and perpendicular to the direction in question.

For example, if the whole pressure of a fluid on the bottom of a vessel is 2000 lbs., and the pressure is uniform throughout, then if we take a square inch as the unit of area, and the area of the bottom of the vessel is 40 square inches, the pressure *at* a point in the base is $\frac{2000}{40}$ lbs. or 50 lbs.

25. The student must carefully observe the distinction between the expressions "pressure *on* a point" and "pressure *at* a point": the former is zero, because a point has no magnitude.

26. If a mass of fluid is at rest, any portion of it may be supposed to become rigid without affecting the conditions of equilibrium.

Thus if we consider any portion A of the fluid in a closed vessel, we may suppose the fluid *in* A to become solid, while the rest of the fluid remains in a fluid state, or we may suppose the fluid *round* A to become solid, while the fluid in A remains in a fluid state.

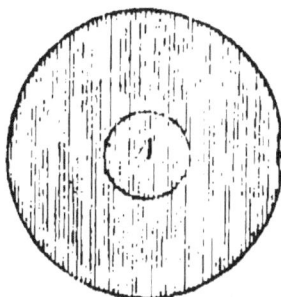

27. The importance of the principle laid down in the preceding article may be seen from the following considerations. The laws of Statics are proved only in the case of forces acting on rigid bodies. Now since the supposition of any part of a fluid becoming solid does not affect the action of the forces acting upon it, and since we can in that case obtain the effect of those forces by the laws of Statics, we shall know their effect on the fluid.

28. If a body of fluid, supposed to be without weight, be confined in a closed vessel, so as to exactly fill the vessel, an equal pressure will be exerted on the fluid by every equal area in the sides of the vessel (Art. 22), and we proceed to shew that the pressure is the same in all directions at every point of the fluid.

For let O be any point in the fluid, and AB, CD two plane surfaces, each representing a unit of area, passing through O

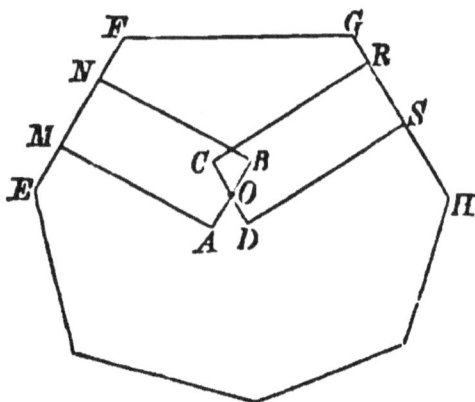

and parallel to two sides of the vessel EF, GH. Then drawing straight lines at right angles to AB, CD from the extromities of AB, CD to the sides of the vessel, we may imagine all the fluid except that contained in the prism $ABNM$ to become solid.

Then the pressure exerted on the fluid by the area MN will be transmitted to AB.

Again, if we suppose all the fluid except that contained in the prism $CDSR$ to become solid, the pressure exerted on the fluid by the area RS will be transmitted to CD.

Now the pressures exerted on the fluid by the areas MN, RS are equal, and consequently the pressures on AB, CD will be equal, that is, the pressure at the point O is the same *in all directions*.

Also since the distance of the point O from the sides of the vessel is not involved in the preceding considerations, it follows that the pressure is the same *at every point*.

CHAPTER II.

On the Pressure of a Fluid acted on by Gravity.

29. In the preceding chapter we considered the consequences that result from the peculiar property, essential to all fluids, of transmitting equally in all directions the pressures applied to their *surfaces.*

We have now to consider the effects produced by the *action of gravity* upon the *substance* of a fluid.

30. The student must mark carefully the distinction between force applied to a surface and force applied to each of the particles composing a body. As an example of these distinct forces consider the case of a book resting on a table. Force is applied to the surface of the book by the table, and thus is counterbalanced the force of gravity which acts upon each particle of which the book is composed.

31. All fluids are subject to the action of gravity in the same way as solid bodies. Each particle of a fluid has a tendency to fall to the surface of the earth, and in a mass of fluid at rest there is a particular point, called the centre of gravity, at which the resultant of all the forces exercised by the attraction of the Earth on the particles composing the fluid may be supposed to act.

32. The term *density* is applied to fluids, as it is to solid bodies, to denote the degree of closeness with which the particles are packed.

When we speak of a fluid of *uniform* density, we mean that if from any part of the body of fluid a portion be taken, and if from any other part of the body of fluid a portion like in form and equal in volume to the former portion be taken, the weights of the two portions will be equal.

33. If a vessel be filled with a heavy fluid of uniform density the pressure at every point in the interior of the fluid will not be the same, because the pressure which results from the action of gravity will vary in magnitude according to the position of the point in the containing vessel.

Consider a closed surface of small dimensions containing the point *A*, and suppose the fluid outside the closed surface to become solid. The fluid *within* the closed surface will exercise pressure against the surface at every point, and these pressures will be unequal, because the fluid is acted on by gravity. But we may conceive that, if the quantity of fluid within the surface be *very small*, the difference between the pressures at different points of the surface will be very small, and when the surface is indefinitely diminished the pressures exercised by the fluid at each point of the surface may be regarded as equal, and the weight of the fluid may be neglected.

Thus we can consider it as the case of a weightless fluid and apply the conclusions of Art. 28.

Hence all the planes of equal area which can be drawn, passing through the point *A* and not extending beyond *the small surface*, may be considered to be subject to equal pressures.

So we conclude that in a heavy fluid of uniform density

(1) The pressure will vary from point to point.

(2) The pressure will be the same in all directions at any particular point.

34. We have next to consider in what way the pressure varies from point to point in the interior of a fluid of uniform density when it is in equilibrium, and first we shall shew *that the pressure is the same at all points in the same horizontal plane.*

Let *A* and *B* be two points in the same horizontal plane in the interior of a fluid of uniform density.

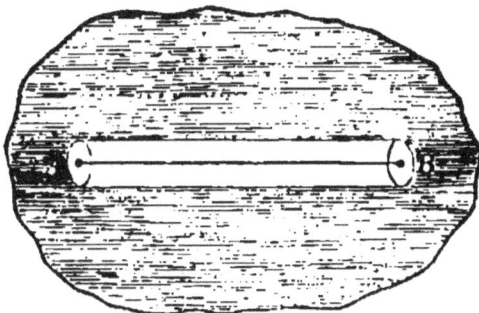

Imagine all the fluid contained in a small horizontal cylinder, of which *AB* is the axis, to become solid.

Then the forces acting on the cylinder are

(1) The fluid pressures on its curved surface ⎱ perpendicular
(2) The weight of the cylinder ⎰ to the axis.

(3) The fluid pressure on the end *A* ⎱ parallel to the axis.
(4) The fluid pressure on the end *B* ⎰

Of these (1) and (2) have no tendency to produce motion in the direction of the axis (Art. 17).

Therefore, since there is no horizontal motion,

fluid pressure on end *A* = fluid pressure on end *B*.

And since, the ends being very small, the pressure at every point in each end may be regarded as the same,

pressure at point *A* = pressure at point *B*.

55. *The pressure at any point within a heavy inelastic fluid, not exposed to external pressure, is proportional to the depth of that point below the surface of the fluid.*

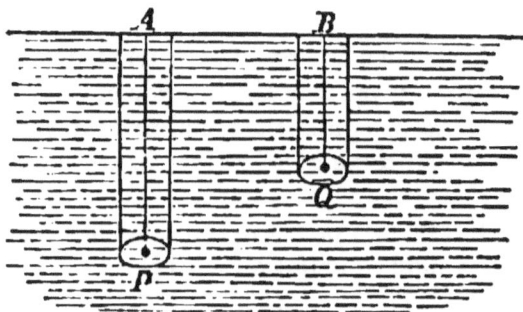

Let P and Q be two points at different depths below the surface of the fluid.

Suppose two small equal and horizontal circles to be described round P and Q as centres.

Then suppose the fluid in the two small vertical cylinders PA, QB, extending from the bases P and Q to the surface, to become solid.

Now the forces acting on the cylinder PA are

(1) The fluid pressures on its curved surface, all of which are perpendicular to the axis.

(2) The weight of the cylinder ⎫
(3) The fluid pressure on the base P ⎬ parallel to the axis.

Of these (1) has no tendency to produce motion in the direction of the axis (Art. 17).

Hence since there is no vertical motion,

> fluid pressure on base P = weight of cylinder PA.

So also, fluid pressure on base Q = weight of cylinder QB.

Hence

pressure at point P : pressure at point Q

> :: pressure on base P : pressure on base Q, (Art. 24.)
> :: weight of cylinder PA : weight of cylinder QB,
> :: length of PA : length of QB (the bases being equal),
> :: depth of P : depth of Q.

Cor. If pressure at P = pressure at Q
> depth of P = depth of Q.

The pressure of the atmosphere on the surface of the fluid is not taken into account, but we shall shew hereafter how it affects the pressure at a point in the interior of a fluid.

36. *The surface of a heavy inelastic fluid at rest is horizontal.*

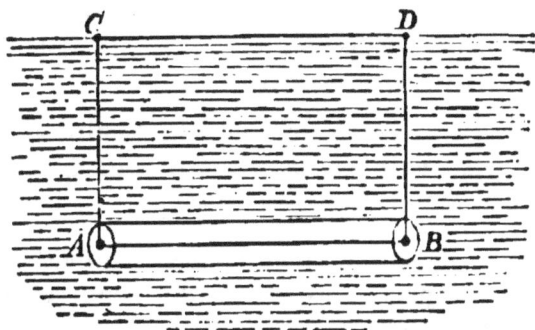

Let *A* and *B* be two points in the same horizontal plane in the interior of a heavy fluid at rest.

Suppose the fluid contained in a small horizontal cylinder of fluid, of which *AB* is the axis, to become solid.

Then, fluid pressure on end *A* = fluid pressure on end *B* (Art. 34), and, since the ends are equal,

fluid pressure at point *A* = fluid pressure at point *B*.

Hence *A* and *B* are at the same depth below the surface of the fluid (Cor. Art. 35), and if we draw *AC*, *BD* vertically to meet the surface in *C*, *D*,
$$AC = BD,$$
also, *AC* is parallel to *BD*;

$$\therefore CD \text{ is parallel to } AB \text{ (Eucl. i. 33)};$$

$$\therefore CD \text{ is horizontal.}$$

Similarly any other point in the surface may be proved to be in the same horizontal plane with *C* or *D*;

$$\therefore \text{ the surface is horizontal.}$$

37. The proposition that the surface of a fluid at rest is horizontal is only true when a very moderate extent of surface is taken.

Large surfaces of water assume, in consequence of the attraction exercised by the earth, a spherical form.

The following practical results are worthy of notice:

(1) All fluids find their level. If tubes of various shapes, some large and some small, some straight and others bent, bo placed in a closed vessel full of water, and water be then poured into one of the tubes, the fluid will rise to a uniform height in it and all the other tubes.

(2) If pipes be laid down from a reservoir to any distance, the fluid will mount to the same height as that to which it is raised in the reservoir.

(3) The surface of a fluid at rest furnishes a means of observing objects at a distance in the same horizontal plane with a mark at the place of observation.

38. We have seen that in an inelastic fluid at rest the pressure at any point depends on the depth of that point below the surface of the fluid, that is, on the length of the vertical line drawn from the point to meet a horizontal line drawn through the highest point in the fluid.

Thus if ABC be a conical vessel with a horizontal base, standing on a table, and filled with fluid, the pressure at any point P is determined in the following manner.

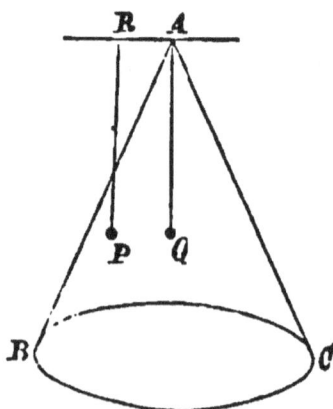

From A, the highest point of the fluid, draw a vertical line meeting the horizontal plane passing through P in the point Q.

Then the pressure at P = pressure at Q, because P and Q are in the same horizontal plane.

But pressure at Q depends on the length of AQ: therefore pressure at P depends on the length of PR, a line drawn vertically to meet the horizontal line AR.

39. *If a vessel, of which the bottom is horizontal and the sides vertical, be filled with fluid, the pressure on the bottom will be equal to the weight of the fluid.*

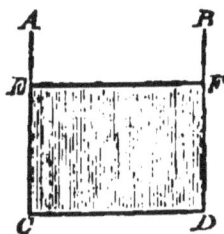 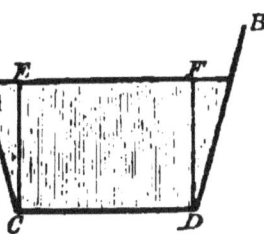 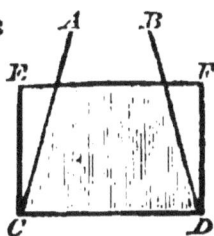

Fig. ɪ.   Fig. ɪɪ.   Fig. ɪɪɪ.

Let *ACDB* (fig. I.) be a vessel whose bottom, *CD*, is horizontal, and its sides vertical. We may consider the fluid in this vessel to be made up of vertical columns of fluid. Each of these columns will press vertically downwards with its weight, and the sum of these pressures will be the weight of the fluid. Now the base of the vessel, being horizontal, will sustain all these vertical pressures;

∴ pressure on the base of the vessel = weight of the fluid.

If the sides of the vessel be not vertical, as in figs. II. and III., the pressure on the base will be equal to the weight of a column of fluid *ECDF*, *EC* and *FD* being perpendicular to *CD*, and *EF* being the surface of the fluid.

Hence if in the three vessels the bases are equal and on the same horizontal plane, and the fluid stands at the same height in the vessels, the pressure on the base in each case will be the same.

The fluid in vessel I. produces a pressure on the base equal to its own weight.

The fluid in vessel II. produces a pressure on the base less than its own weight.

The fluid in vessel III. produces a pressure on the base greater than its own weight.

Examples.—II.

(1) If the pressure at a depth of 32 feet be 15 lbs. to the square inch, what will the pressure be at a depth of 42 feet 6 inches?

(2) If the pressure at a depth of 8 feet be 14¾ lbs. to the square inch, what will be the pressure at a depth of 20 ft. 6 in.?

(3) In two uniform fluids the pressures are the same at the depths of 3 and 4 inches respectively: compare the pressures at the depths of 7 and 8 inches respectively.

(4) In two uniform fluids the pressures are the same at the depths of 2 and 3 inches respectively: compare the pressures at the depths of 9 and 12 inches respectively.

(5) Find the height of a column standing in water 30 feet deep, when the pressure at the bottom is to the pressure at the top as 3 to 2.

(6) If the pressure of a uniform fluid, not exposed to external pressure, be 15 lbs. to the square inch at a depth of 15 feet, what will be the pressure at a depth of 12 feet?

(7) If the pressure of a uniform fluid, not exposed to external pressure, be 3 lbs. to the square inch at a depth of 4 feet, what will be the pressure on a square inch at a depth of 12 feet?

(8) What is the pressure on the horizontal bottom of a vessel filled with water to the depth of 2½ feet, the area of the base being 20 square feet, and the weight of a cubic foot of water 1000 oz.?

(9) A cubic foot of mercury weighs 13600 oz. Find the pressure on the horizontal base of a vessel containing mercury, the area of the base being 8 square inches, and the depth of the mercury 3 inches.

(10) What is the pressure on the horizontal base of a vessel filled with water to the depth of 15 feet, the area of the base being 24 square feet, and the weight of a cubic foot of water 1000 oz.?

(11) A cistern shaped like an equilateral triangle of which one side is 6 feet is filled with water to the depth of two feet: find the pressure on the base, the weight of a cubic foot of water being 1000 oz.

(12)   The spout of a teapot springs from the middle point of one side, and its upper extremity is on a level with the lid. If the spout be broken off half-way, how high can the teapot be filled ?

(13)   When bottles that have been sunk in deep water have been brought up, their corks have been found driven in.   How do you explain this ?

(14)   If a pipe, whose height above the bottom of a vessel is 112 feet, be inserted vertically in the vessel, and the whole be filled with water, find the pressure in tons on the bottom of the vessel, the area of the bottom being 4 square feet, and the weight of a cubic foot of water 1000 oz.

(15)   A hole, a square inch in area, is bored in the flat cover of a vessel full of water, and a smooth piston weighing 7 lbs. 13 oz. is fitted into it; a vertical tube is then fitted into another hole in the cover, and water is poured into it: find how high the water must be made to ascend in it in order that the piston may be driven out, a cubic foot of water weighing 1000 oz.

CHAPTER III.

On Specific Gravity.

40. Some substances are from the nature of their composition more weighty than others. We call gold a heavier metal than lead, because we know by experience that a given volume of gold is more weighty than an equal volume of lead.

41. We make a distinction between the terms *weight* and *weightiness*.

We speak of the weight of a particular lump of gold or iron.

We speak of the weightiness of gold or iron, not referring to any particular lump, but to the special characteristics of the metals in question.

Further we say that gold is heavier than iron, having no particular lump of the metals in view, but expressing our notions of the degree of weightiness that is peculiar to either substance.

This degree of weightiness is known by the name Specific Gravity.

DEF. *The Specific Gravity of a substance is the degree of weightiness of that substance.*

42. If of two substances, one of which is twice as weighty as the other, we take two lumps of equal volume, the weight of one lump is evidently twice that of the other: and, generally, if one substance be S times as weighty as the other, the weight of any volume of the first is S times the weight of an equal volume of the other. Now by a substance, the measure of the specific gravity of which is S, we mean a substance which is S times as weighty as the standard by which specific gravities are estimated. Therefore any volume of this substance will weigh S times as much as the equal volume of the standard.

43. The requisites for a Standard are that it should be definite and uniform, and these requisites are possessed by Pure Distilled Water at a certain temperature. This substance is therefore taken as the standard for estimating the specific gravities of solid bodies and inelastic fluids.

44. When we say that the specific gravity of gold is 19, we mean that the specific gravity of gold is 19 times that of Pure Distilled Water, and therefore a given volume of gold weighs 19 times as much as the same volume of distilled water.

45. To measure the Weight of a body we must have a unit of weight, and to measure the Volume of a body we must have a unit of volume. These units we may select in any way that may suit our purpose, and we connect them with the unit of specific gravity by the following convention :

The unit of specific gravity is the specific gravity of that substance of which a unit of volume contains a unit of weight.

46. *To find the numerical relation existing between the measure of the specific gravity of a substance and the measures of the weight and volume of any given quantity of the substance.*

Let W represent the measure of the weight of a substance, that is the number of times it contains the unit of weight.

Also, let V represent the measure of the volume of the substance, that is the number of times it contains the unit of volume.

And let S represent the measure of the specific gravity of the substance, that is the number of times it contains the unit of specific gravity.

Then one unit of volume of this substance will weigh S times as much as a unit of volume of the standard substance, 'Art. 42) that is, its weight is S times the unit of weight.

Therefore the weight of V units of volume is VS times the unit of weight;

therefore the measure of the weight of V units of volume of the substance is VS;

but this measure we have denoted by W;

$\therefore W = VS.$

47. The equation $W = VS$ gives us merely the relation between three numbers, and two of these must be given in order that we may determine the third.

When we have found it we know *the measure* of the weight or volume or specific gravity, as the case may be, and we must have the unit of weight, or of volume, or of specific gravity also given to enable us to determine the weight or volume or specific gravity of a particular substance. So that we may put it thus :

$$\text{measure of weight} = VS,$$

$$\text{measure of volume} = \frac{W}{S},$$

$$\text{measure of specific gravity} = \frac{W}{V} ;$$

and

$$\text{weight} = VS \text{ times (unit of weight)},$$

$$\text{volume} = \frac{W}{S} \text{ times (unit of volume)},$$

$$\text{specific gravity} = \frac{W}{V} \text{ times (unit of specific gravity)}.$$

48. A cubic foot of pure distilled water at a temperature of 62° Fahrenheit weighs about 998 oz., and for rough calculations it is assumed that the weight of a cubic foot of water is 1000 ounces.

Then if we take 1 cubic foot as our unit of volume and pure distilled water as our standard of specific gravity, the unit of weight will be 1000 ounces.

Or if we prefer to take 1 lb. avoirdupois as our unit of weight and pure distilled water as our standard of specific gravity, the unit of volume will be $\frac{16}{1000}$ of a cubic foot, that is ·016 cub. ft.

49. We shall next explain how quantities are measured; and then we shall give three examples, worked out first on the supposition that 1 cubic foot is taken as the unit of volume, and secondly, on the supposition that 1 lb. avoirdupois is taken as the unit of weight, so that the student may see that the same result must follow from both suppositions, and that such a choice may be made as to the units as may be suitable to any particular case.

50. To *measure* any quantity we fix upon some definite quantity of the same kind for our standard, or *unit*, and then any quantity of that kind is measured by finding how many times it contains this unit, and this number of times is called the *measure* of the quantity.

For example, if one pound avoirdupois be the unit of weight, the measure of 16 lbs is 16. Or, to put our calculations in a tabular form, we may give the following Examples:

Unit.	Quantity.	Measure.
1 lb. avoird.	8 lbs.	8.
1 lb. avoird.	4 oz.	$\dfrac{1}{4}$.
1 lb. avoird.	1 lb. troy.	$\dfrac{5760}{7000}$.
1 cub. ft.	$6\frac{1}{2}$ cub. ft.	6·5.
1 cub. ft.	3 cub. in.	$\dfrac{3}{1728}$.
1000 oz. av.	14 lbs. av.	$\dfrac{14 \times 16}{1000}$.
·016 cub. ft.	5 cub. in.	$\dfrac{5}{1728 \times ·016}$.

51. First, when 1 cubic foot is taken as the unit of volume, and consequently 1000 oz. as the unit of weight, to solve the following examples :

Ex. (1) The specific gravity of lead is 11·4, find the weight of 720 cubic inches of lead.

Here $V = \dfrac{720}{1728}$, $S = 11\cdot4$.

Weight required $= VS$ (unit of weight)

$$= \left(\dfrac{720}{1728} \times 11\cdot4\right) \text{ times } 1000 \text{ oz.}$$

$$= 4750 \text{ oz.}$$

$$= 296\dfrac{7}{8} \text{ lbs.}$$

Ex. (2) If 5 cubic feet of a substance weigh 240 lbs., what is its specific gravity?

Here $W = \dfrac{240 \times 16}{1000}$, $V = 5$.

Sp. gr. required $= \dfrac{W}{V}$ (unit of specific gravity)

$$= \dfrac{\dfrac{240 \times 16}{1000}}{5} \text{ (unit of specific gravity)}$$

$$= \dfrac{240 \times 16}{1000 \times 5} \text{ (unit of specific gravity)}$$

$$= \cdot768 \text{ (unit of specific gravity).}$$

Ex. (3) What is the volume of a substance whose specific gravity is 9·6 and whose weight is 4200 lbs. ?

Here $W = \dfrac{4200 \times 16}{1000}$, $S = 9\cdot6$.

Volume required $= \dfrac{W}{S}$ (unit of volume)

$$= \dfrac{\dfrac{4200 \times 16}{1000}}{9\cdot6} \text{ cub. ft.}$$

$$= 7 \text{ cub. ft.}$$

52. Secondly, when 1 lb. avoirdupois is taken as the unit of weight, and consequently ·016 cub. ft. as the unit of volume, our examples will stand thus:

Ex. (1)

Here
$$V=\frac{720}{1728 \times ·016}, \ S=11·4.$$

Weight required $= VS$ (unit of weight)

$$=\left(\frac{720}{1728} \times \frac{1}{·016} \times 11·4\right) \text{ times 1 lb.}$$

$$=296\frac{7}{8} \text{ lbs.}$$

Ex. (2)

Here
$$W=240, \ V=\frac{5}{·016}.$$

Sp. gr. required $= \dfrac{W}{V}$ (unit of specific gravity)

$$=\frac{240}{\dfrac{5}{·016}} \text{ (unit of specific gravity)}$$

$$=\frac{240 \times ·016}{5} \text{ (unit of specific gravity)}$$

$$=·768 \text{ (unit of specific gravity).}$$

Ex. (3)

Here
$$W=4200, \ S=9·6.$$

Volume required $= \dfrac{W}{S}$ (unit of volume)

$$=\frac{4200}{9·6} \text{ times ·016 cub. ft.}$$

$$=\frac{4200 \times 16}{9·6 \times 1000} \text{ cub. ft.}$$

$$=7 \text{ cub. ft.}$$

53. If a number of substances be put together to form a mixture, we shall *generally* have the following relations :

(1) sum of measures of weights of compounds = measure of weight of mixture.

(2) sum of measures of volumes of compounds = measure of volume of mixture.

Thus if $w_1, w_2, w_3,\ldots\ldots$ be the measures of the weights,

$$v_1, v_2, v_3,\ldots\ldots\ldots\ldots\ldots\ldots\ldots\ldots\ldots \text{ volumes,}$$

$$s_1, s_2, s_3,\ldots\ldots\ldots\ldots\ldots\ldots\ldots\ldots \text{ specific gra-}$$

vities of the compounds, and

w, v, s the measures of the weight, volume and specific gravity of the mixture, we shall have

$$w_1 + w_2 + w_3 + \ldots\ldots\ldots\ldots = w,$$

$$v_1 + v_2 + v_3 + \ldots\ldots\ldots\ldots = v ;$$

and therefore

$$v_1 s_1 + v_2 s_2 + v_3 s_3 + \ldots\ldots\ldots = vs,$$

$$\frac{w_1}{s_1} + \frac{w_2}{s_2} + \frac{w_3}{s_3} + \ldots\ldots\ldots = \frac{w}{s}.$$

NOTE. We say that these relations hold *generally*, because in some cases, when substances are mixed, the volume of the mixture is not equal to the sum of the volumes of the two substances. For instance, 70 pints of sulphuric acid mixed with 30 pints of water will make a mixture of less than 99 pints.

54. In applying these formulæ to the solution of examples, we may take any unit of volume or of weight, adhering to it through the whole calculation.

Ex. (1) To find the specific gravity of a mixed metal composed of 5 cubic inches of copper, specific gravity 9, and 8 cubic inches of tin, specific gravity 7·2.

Since $v_1 s_1 + v_2 s_2 = vs$,

if we take 1 cubic inch as the unit of volume, we have

$$5 \times 9 + 8 \times 7\cdot2 = (5+8) s;$$

$$\therefore s = \frac{45 + 57\cdot6}{13} = 7\cdot96 \text{ nearly.}$$

Ex. (2) Ten pounds of fluid, specific gravity 1·05, are mixed with 15 pounds of distilled water. Find the specific gravity of the mixture·

Since
$$\frac{w_1}{s_1} + \frac{w_2}{s_2} = \frac{w}{s},$$

if we take 1 lb. as the unit of weight, we have

$$\frac{10}{1\cdot05} + \frac{15}{1} = \frac{25}{s};$$

$$\therefore \frac{2}{1\cdot05} + 3 = \frac{5}{s};$$

$$\therefore \frac{5}{s} = \frac{5\cdot15}{1\cdot05};$$

$$\therefore s = \frac{105 \times 5}{515} = \frac{105}{103} = 1\cdot019 \text{ nearly.}$$

55. *The Density of a substance is the degree of closeness with which the particles composing the substance are packed together.*

The difference between density and specific gravity may be stated thus: in estimating the density of a body we take into account the quantity of matter contained in a given volume: in estimating the specific gravity of a body we take into account the effect of the action of gravity on a given volume.

If we take the same substance, as pure distilled water, as that to which we refer as a standard in measuring the density and specific gravity of another substance, the measures of the density and specific gravity will be the same.

EXAMPLES.—III.

(1) THE specific gravity of copper is 8·91; find the weight of 512 cubic inches of copper, a cubic foot of water weighing 1000 oz.

(2) If 4 cubic inches of iron weigh as much as 72 cubic inches of amber, compare the specific gravities of iron and amber.

(3) The specific gravity of mercury being 13·5, find the weight of one cubic inch of it, having given that a cubic foot of water weighs 1000 oz.

(4) If two cubic foot of a substance weigh 100 lbs., what is its specific gravity ?

(5) Find the weight of 36 cubic inches of cork, whose specific gravity is 0·24.

(6) A cubic foot of water weighs 1000 oz., what will be the weight of a cubic inch of a substance whose specific gravity is 3 ?

(7) What is the specific gravity of a body of which m cubic feet weigh n lbs. ?

(8) Five cubic inches of iron weigh 22½ oz., what is the specific gravity of iron?

(9) Twelve cubic feet of dried oak weigh 875 lbs., what is the specific gravity of the wood ?

(10) Twenty-six cubic feet of ash weigh 1371½ lbs., what is its specific gravity ?

(11) A metal, whose specific gravity is 15, is mixed with half the volume of an alloy whose specific gravity is 12, find the specific gravity of the compound.

(12) Two metals are combined into a lump the volume of which is 2 cubic inches ; 1½ cubic inches of one metal weigh as much as the lump, and 2½ cubic inches of the other metal weigh the same. What volume of each of the two metals is there in the lump ?

(13) Two substances whose specific gravities are 1·5 and 3·0 are mixed together, and form a compound whose specific gravity is 2·5; compare the volumes and also the weights of the two substances.

(14) The specific gravity of sea-water being 1·027, what proportion of fresh water must be added to a quantity of sea-water that the specific gravity of the compound may be 1·009 ?

(15) Equal weights of two substances whose densities are 3·25 and 2·75 are mixed together; find the density of the compound.

(16) Equal volumes of two substances whose specific gravities are 2·5 and 1·5 are mixed together; what is the specific gravity of the compound?

(17) Five cubic inches of lead, specific gravity 11·35, are mixed with the same volume of tin, specific gravity 7·3; what is the specific gravity of the compound?

(18) A mixture is formed of equal volumes of three fluids; the densities of two are given and also the density of the mixture. What is the density of the third fluid?

(19) Ten cubic inches of copper, specific gravity 8·9, are mixed with seven cubic inches of tin, specific gravity 7·3; find the specific gravity of the compound.

(20) Three fluids, whose specific gravities are ·7, ·8 and ·9 respectively, are mixed in the proportion of 5 lbs., 6 lbs., and 7 lbs. What is the specific gravity of the mixture?

(21) The specific gravity of pure gold is 19·3 and of copper 8·62 ; required the specific gravity of standard gold, which is a mixture of eleven parts of gold and one of copper.

(22) When 63 pints of sulphuric acid, specific gravity 1·82, are mixed with 24 pints of water, the mixture contains only 86 pints. What is its specific gravity?

(23) If three fluids the volumes of which are 4, 5, 6 and the specific gravities 2, 3, 4 are mixed together, determine the specific gravity of the compound.

(24) The specific gravity of quartz is 2·62, and that of gold 19·35 ; a nugget of quartz and gold weighs 11·5 oz., and its specific gravity is 7·43; find the weight of gold in it.

(25) An iron spoon is gilded, and the mean specific gravity of the gilded spoon is 8; those of iron and gold are 7·8 and 19·4 : find the ratio of the volumes and weights of the metals employed.

CHAPTER IV.

On the Conditions of Equilibrium of Bodies under the Action of Fluids.

56. WHEN a body is wholly or partially immersed in a fluid, it is a general principle of Hydrostatics that *the resultant pressure of the fluid on the surface of the body is equal to the weight of the fluid displaced.* This principle we shall prove for two cases in Articles 57 and 61.

(1) When the body is *wholly* immersed in the fluid:

(2) When the body is *partially* immersed in the fluid.

57. *To find the resultant Pressure of a Fluid on a body wholly immersed and floating in a fluid.*

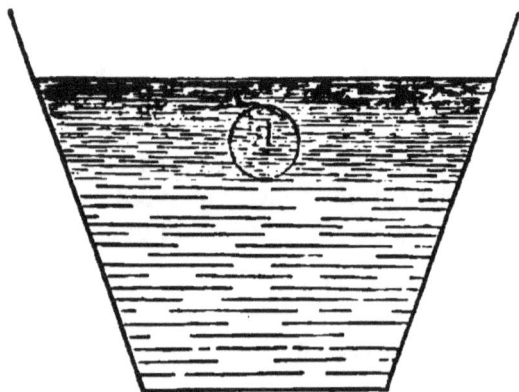

Let *A* be a body floating in a fluid and wholly immersed in it.

Imagine the body removed and the vacant space filled with fluid of the same kind as that in which the body floated.

Then suppose this substituted fluid to become solid.

. The pressure at each point of its surface will still be the same as it was at the same point of the surface of *A*

The solidified fluid is kept at rest by

(1) The attractions exercised by the earth on every particle of its mass :

(2) The pressures exercised by the fluid at the different points of its surface.

Hence the resultants of these two sets of forces must be *equal in magnitude* and *opposite in their lines of action.*

Now the resultant of set (1) is called the weight of the solidified fluid and acts vertical.y *downwards* through its centre of gravity.

Hence the resultant of set (2) is equal in magnitude to the weight of the solidified fluid and acts vertically *upwards* through its centre of gravity.

Now since the pressures on the solidified fluid are the same as on the body *A*, we see that the resultant pressure of the fluid on *A* is equal to the weight of the fluid displaced by *A* and acts vertically upwards through the centre of gravity of this displaced fluid.

This principle we shall now apply to the following Examples in Statics.

58. Ex. I. *Find the conditions of equilibrium of a body floating in a fluid and wholly immersed in it.*

The body *A* (see diagram in Art. 57) is kept at rest by

(1) Its weight, acting vertically downwards through its centre of gravity:

(2) The pressures of the fluid on its surface, the resultant of which is equal to the weight of the fluid displaced by *A* and acts vertically upwards through the centre of gravity of the fluid displaced.

Hence

(1) Weight of A = weight of fluid displaced by A :

(2) The centres of gravity of A and of the fluid displaced are in the same vertical line.

These are the conditions of equilibrium.

Note. A difficulty often occurs with beginners in conceiving how a solid body can be in equilibrium *in the midst of a fluid*, neither rising to the surface nor sinking to the bottom. It may however be proved by experiment that a hollow ball of copper, such as is used for a ball-tap, may be constructed of such a weight relatively to its size that when placed in water it will remain where it is placed, just as the body A is represented in the diagram.

59. **Ex. II.** *Find the conditions of equilibrium for a body of uniform density wholly immersed in a fluid and in part supported by a string.*

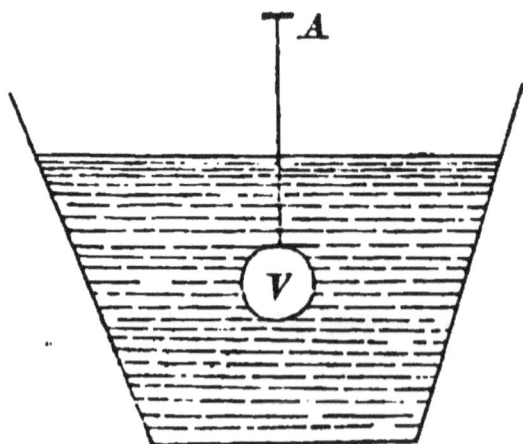

Let a body the measure of whose volume is V be suspended by a string from the fixed point A so as to float below the surface of a fluid.

The body is kept at rest by

(1) its weight,

(2) the pressures of the fluid on its surface,

(3) the tension of the string.

Now (1) is equivalent to a single resultant acting vertically *downwards* through the centre of gravity of the body;

  (2) is equivalent (by Art. 57) to a single resultant, equal to the weight of fluid displaced and acting vertically *upwards* through the centre of gravity of the fluid displaced:

(and these two centres of gravity coinciding)

therefore (3) must act (see Statics, Art. 52) *upwards* in the vertical line through this common centre of gravity,

and (1) must be equal to the sum of (2) and (3).

Hence, if

S be the measure of the specific gravity of the body,

S' .... of the fluid, ·

T .... of the tension of the string,

there is equilibrium when

$$VS = VS' + T$$

or $T = V(S - S')$.

Ex. A piece of metal, whose specific gravity is 7·3 and volume 24 cubic inches, is suspended by a string so as to be wholly immersed in water. Find the tension of the string.

Taking 1 cubic inch as the unit of volume, and consequently $\dfrac{1000}{1728}$ oz. as the unit of weight,

$$\text{tension of string} = 24\,(7\text{·}3 - 1) \times \frac{1000}{1728} \text{ oz.}$$

$$= \frac{24 \times 6\text{·}3 \times 1000}{1728} \text{ oz.}$$

$$= 875 \text{ oz.}$$

60. Ex. (3) *If a body of uniform density be immersed in a fluid and be prevented from rising by a string attached to the bottom of the vessel containing the fluid, find the tension of the string.*

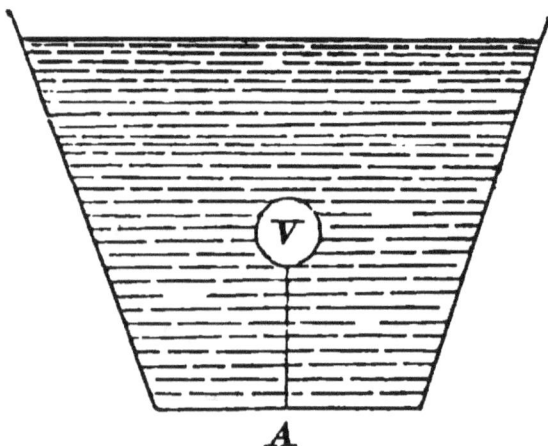

Let a body, the measure of whose volume is V, be kept under the surface of a fluid by a string fastened to A, a point in the base of the vessel.

The body is kept at rest by

(1)   its weight, acting vertically downwards,

(2)   the tension of the string, acting vertically downwards,

(3)   the resultant of fluid pressures on the body, acting vertically upwards.

Hence, if

T be the measure of the tension of the string,

S .... specific gravity of the body,

S' .... specific gravity of the fluid,

since there is equilibrium,

$$VS + T = VS';$$

$$\therefore T = VS' - VS$$

$$= V(S' - S).$$

61. *To find the resultant pressure of a fluid on a body partially immersed and floating in the fluid.*

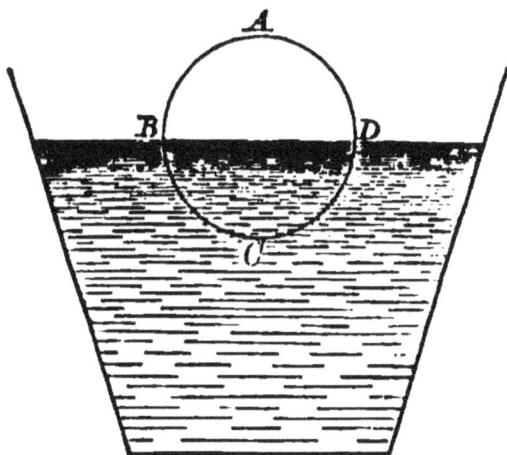

Let *ABCD* be a body partially immersed and floating in a fluid, the part *BCD* being below the surface of the fluid.

Imagine the body removed and the vacant space *BCD* filled with fluid of the same kind as that in which the body floated.

Then suppose this substituted fluid to become solid.

The pressure at each point of its surface will still be the same as it was at the same point of *BCD*.

The solidified fluid is kept at rest by

    (1) the attractions exercised by the Earth on every particle of its mass,

    (2) the pressures exercised by the fluid at the different points of its surface*.

Hence the resultants of these two sets of forces must be *equal in magnitude* and *opposite in their lines of action.*

---

* Throughout this chapter the space occupied by the air is supposed to be a vacuum.

Now the resultant of set (1) is called the weight of the solidified fluid, and acts vertically *downwards* through its centre of gravity.

Hence the resultant of set (2) is equal in magnitude to the weight of the solidified fluid, and acts vertically *upwards* through its centre of gravity.

Now since the pressures on the solidified fluid are the same as on the surface *BCD*, we see that the resultant pressure of the fluid on the floating body is equal to the weight of the fluid displaced, and acts vertically upwards through the centre of gravity of the displaced fluid.

This principle we shall now apply to the following examples in Statics.

62. Ex. I. *Find the conditions of equilibrium of a body floating and partially immersed in a fluid of uniform density.*

The body *ABCD* (see diagram in Art. 63) is kept at rest by

(1)   its weight acting vertically downwards through its centre of gravity,

(2)   the pressures of the fluid on the surface *BCD*, the resultant of which is equal to the weight of fluid displaced by the body, and acts vertically upwards through the centre of gravity of the fluid displaced.

Hence

(1)   weight of the body = weight of fluid displaced;

(2)   the centres of gravity of the body and of the fluid displaced are in the same vertical line.

These are the conditions of equilibrium.

63. **Ex. II.** *When a body of uniform density floats in a fluid, the volume of the part immersed is to the volume of the whole body as the specific gravity of the body is to the specific gravity of the fluid.*

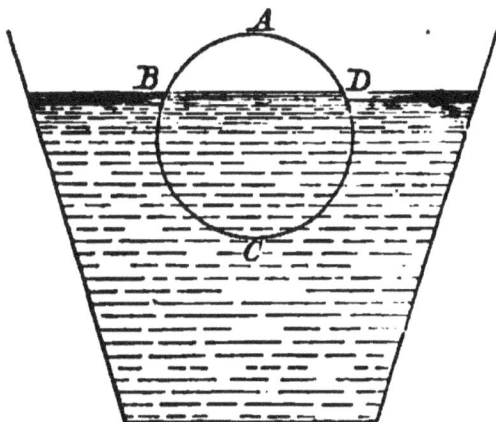

Let V be the measure of the volume of the whole body $ABCD$,

V' .... part immersed BCD.

S .... specific gravity of the body,

S' .... specific gravity of the fluid.

Then since, Art. 62,

weight of floating body = weight of displaced fluid,

$$V.S = V'.S';$$

$$\therefore V' : V :: S : S'.$$

Ex. A solid, whose specific gravity is ·4, floats in a fluid whose specific gravity is 1·2. What part of the solid is below the surface?

Let x be the measure of the part immersed,

m the measure of the whole body.

Then

$$x : m = ·4 : 1·2;$$

$$\therefore x = \frac{·4}{1·2} m = \frac{4}{12} m = \frac{1}{3} m.$$

64. *The Hydrostatic Balance.*

The *Hydrostatic Balance* is a common balance with a hook attached to the bottom of one of the scales from which a solid may be suspended and weighed successively (1) in air and (2) when immersed in a fluid.

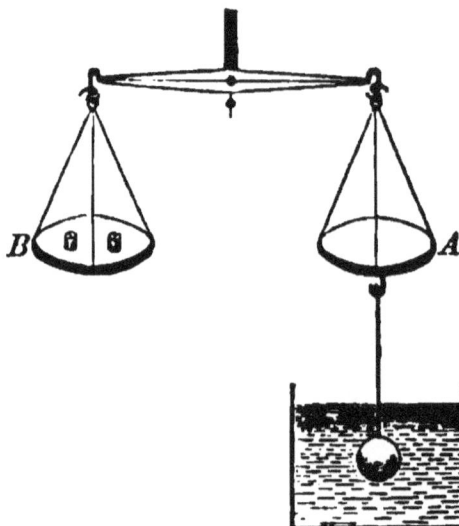

Call the scale to which the hook is attached A and the other scale B. Then by the weight of the solid in air we mean the weight which when placed in B balances the solid suspended in air from A.

And by the weight of the solid in the fluid we mean the weight which when placed in B balances the solid suspended from A so as to be immersed in the fluid.

The difference between these weights is caused by the pressures of the fluid on the surface of the solid, the resultant of these pressures being a force acting vertically upwards and *equivalent to the weight of the fluid displaced by the solid.*

Now if V be the measure of the volume of the solid,

S'........................... specific gravity of the fluid,

measure of weight of fluid displaced by the solid $= VS'$.

65. *To compare the specific gravities of a solid and a fluid by means of the Hydrostatic Balance.*

Let V be the measure of the volume of the solid,

S .... specific gravity of the solid,

S'........................... specific gravity of the fluid,

W .. .. weight of the solid in air.

CASE I. *When the solid is of greater specific gravity than the fluid.*

Let W' be the measure of the weight of the solid in the fluid, then $W - W' =$ the measure of the weight of fluid displaced by the solid,

$$= VS'.$$

Also
$$W = VS ;$$

$$\therefore \frac{VS}{VS'} = \frac{W}{W - W'},$$

or,
$$\frac{S}{S'} = \frac{W}{W - W'},$$

and thus S and S' may be compared.

CASE II. *When the solid is of less specific gravity than the fluid.*

Attach to the solid some heavy substance, called the sinker, which will make the solid sink with it in the fluid.

Let w be the measure of the weight of the two bodies in air,

x .... in the fluid,

y .... sinker in air,

z .... in the fluid.

Then

$w - x =$ measure of weight of fluid displaced by the two bodies,

$y - z =$ .... the sinker.

Subtracting,

$w - x - y + z =$ measure of weight of fluid displaced by the solid

$$= VS';$$

also
$$W = VS;$$

$$\therefore \frac{w - x - y + z}{W} = \frac{S'}{S},$$

and thus S and S' may be compared.

66.   *The common Hydrometer.*

The common Hydrometer consists of a straight stem AB terminating in two hollow spheres C and D. D is usually loaded with mercury, so that the instrument may float in a fluid with the stem vertical.

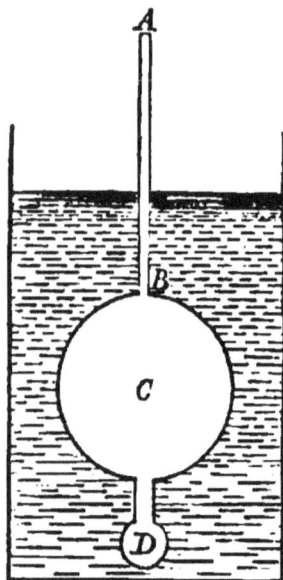

The instrument is used in comparing the specific gravities of two fluids. The stem is marked with graduations by means of which it can be seen what part of the instrument is below the surface of the fluid in which it floats.

When the instrument is placed in a fluid the measure of whose specific gravity is S, suppose that the measure of the bulk of the part immersed is V.

When the instrument is placed in a fluid the measure of whose specific gravity is S', suppose that the measure of the bulk of the part immersed is V'.

Then weight of hydrometer = weight of first fluid displaced
$$= V. S \text{ times the unit of weight ;}$$

and weight of hydrometer = weight of second fluid displaced
$$= V'. S' \text{ times the unit of weight :}$$

$$\therefore V. S = V'. S' ;$$

$$\therefore S : S' :: V' : V;$$

and thus S and S' may be compared, since V and V' are known from the graduations.

67. *Nicholson's Hydrometer.*

This instrument consists of a hollow cylinder of copper A, from which a slender steel wire rises, supporting a dish C. An iron stirrup fixed to the lower end of A supports a heavy dish D. A fine well-defined mark is placed at some point B on the steel wire.

This instrument is used for two purposes :—

(1) To compare the specific gravities of a solid and a fluid.

Let W be the measure of the weight which placed in C causes the hydrometer to sink in the fluid till the surface of the fluid meets the steel wire in B.

Place the solid in C and let X be the measure of the weight added to make the instrument sink to B.

Place the solid in D and let Y be the measure of the weight placed in C to make the instrument sink to B.

Then measure of weight of solid in air $= W - X$,

..........................in the fluid $= W - Y$,

$\therefore$ measure of weight of fluid displaced by solid
$$= (W - X) - (W - Y)$$
$$= Y - X;$$

$$\therefore \frac{\text{S. G. of solid}}{\text{S. G. of fluid}} = \frac{W - X}{Y - X}.$$

(2) To compare the specific gravities of two fluids.

Let W be the measure of the weight of the hydrometer, x and y the measures of weight to be placed in C to make the instrument sink to B in each fluid.

The measure of weight of first fluid displaced $= W + x$,

.... second .... $= W + y$,

and, since the volume is the same in both cases,

$$\frac{\text{S. G. of first fluid}}{\text{S. G. of second fluid}} = \frac{W + x}{W + y}.$$

68. *To compare the specific gravities of two fluids by weighing the same solid in each.*

Let S and S' be the measures of specific gravities of the fluids,

w and w' the measures of weights of the solid when immersed in the respective fluids,

W the measure of weight of the solid in air.

Then $W - w =$ measure of weight of fluid displaced by solid in one case,

$W - w' =$ measure of weight of fluid displaced by solid in the other case ;

$$\therefore \frac{S}{S'} = \frac{W - w}{W - w'},$$

and thus S and S' may be compared.

EXAMPLES.—IV.

(1) A piece of glass when weighed in water loses $\frac{3}{10}$ths of its weight ; what is its specific gravity?

(2) Find the pressure on 28 miles of a submarine telegraphic cable whose circumference is 3 inches, the depth of the cable below the surface of the sea being 480 feet, and the specific gravity of sea water 1·026.

(3) A body whose specific gravity is 3·3 floats on a fluid whose specific gravity is 4·4 ; what portion of the body will be immersed ?

(4) If the specific gravity of standard gold be 19·4, and the weight of a sovereign in air be 5 dwts. $2\frac{1}{2}$ grs., find its weight in water.

(5) If a substance weigh 8 lbs. in air and 6 lbs. in water, what is its specific gravity ?

(6) A cylindrical tub of given weight floats with one-fourth of its axis below the surface of a fluid : find the least weight which will totally immerse the tub.

(7) A body whose specific gravity is 1·4 floats in a fluid whose specific gravity is 2·1; what portion of the body is immersed ?

(8) A leaden bullet, weighing 1 oz., is placed in a glass of water standing on a table ; find the pressure of the bullet on the bottom of the glass, the specific gravity of lead being 11·4.

(9) A cubic inch of cork floats in water ; find the weight which must be placed upon it to cause the half of it to be immersed, the specific gravity of cork being ·24, and the weight of a cubic foot of water 1000 oz.

(10) A cork, whose weight is 1 oz. and specific gravity ·25, is attached by a string to the bottom of a vessel containing water so that the cork is wholly immersed. What is the tension of the string ?

(11) A person supports a ball of lead, weighing 46 oz. and of specific gravity 11·5, wholly immersed in water, by holding the end of a string attached to the ball. What is the tension of the string ?

(12) A vessel containing water is placed in one scale of a balance and weighs 1 lb. A piece of wood of specific gravity ·24 and volume 1 inch is attached to the bottom so as to be immersed. What weight will now balance the vessel ?

(13) A cube hanging by a string is half immersed in water. If the weight of the cube be a pound, and its specific gravity three times that of water, what will be the tension of the string ?

(14) A certain substance weighs 30 oz. in water, and 42 oz. out of water. What is its specific gravity ?

(15) A substance weighs 14 lbs. in water and 2560 oz. out of water. What is its specific gravity ?

(16) A substance weighs 12 oz. in air : a substance weighing 20 oz. in water is attached to it, and the two together weigh 18 oz. in water. What is the specific gravity of the former substance ?

(17) A piece of mahogany weighs in air 375 grains, a piece of brass weighing 380 grains in water is attached to it, and the two together weigh in water 300 grains. What is the specific gravity of the mahogany ?

(18) A piece of metal weighs 113 grains in water and 120 grains in air. What is its specific gravity ?

(19) A piece of calcareous spar weighs in air 190 grains and in water 120 grains. Find its specific gravity.

(20) A body weighs 4 oz. in vacuo, and if another body which weighs 3 oz. in water be attached to it the two together weigh in water 2¼ oz. Find the specific gravity of the former body.

(21) A piece of wood weighs 12 lbs., and when attached to 22 lbs. of lead and immersed in water the two together weigh 8 lbs. If the specific gravity of lead be 11·35 find the specific gravity of the wood.

(22) If the sinker be equal in magnitude to the substance whose specific gravity is required, but double its weight in vacuo, and if the two together weighed in water would balance the sinker in vacuo, what is the specific gravity of the substance ?

(23) The specific gravity of cork is ·24, and the weight of a cubic foot of water is 1000 oz.; find the pressure necessary to hold down under water a cubic foot of cork.

(24) A cylinder floats vertically in a fluid with 8 feet of its length above the fluid; find the whole length of the cylinder, the specific gravity of the fluid being three times that of the cylinder.

(25) A cylinder floats with ⅓th of its bulk above the surface of a fluid whose specific gravity is ·825, find the specific gravity of the cylinder.

(26) Why is it easier to swim in salt water than in fresh ?

(27) Water is poured into a vessel containing mercury, and an iron cylinder allowed to sink through the water floats with its axis vertical in the mercury. If the cylinder be 1 inch in length, find the length of the portion immersed in the mercury. The specific gravity of iron is 7·8, and that of mercury 13·6.

(28) A body, whose specific gravity is ·5, floats on water; if the weight of the body be 1000 oz., find the number of cubic inches of it above the surface of the fluid.

(29) A body containing 12 cubic inches weighs in air 8 lbs.; determine its weight in water.

(30) If a cube float on water with one face horizontal, and a body weighing $\dfrac{1000}{3}$ oz., when placed upon it, make it sink through an inch, find the size of the cube : a cubic foot of water weighing 1000 oz.

(31) What is the specific gravity of a substance, if a hollow rectangular box, ten inches long, eight inches wide, six inches deep, and a quarter of an inch thick, if made of this substance, will just float in water ?

(32) A lamina in the form of an equilateral triangle floats on a fluid with one of its sides horizontal and its vertex downwards. If the density of the triangle be one-third that of the fluid, find the depth of its vertex below the surface.

(33) A triangular lamina of uniform thickness floats in a vertical position with its base horizontal and its sides half immersed in a fluid : compare the specific gravity of the lamina with that of the fluid.

(34) A symmetrical body, weighing 8 lbs., with a weight on the top floats just immersed in a fluid: how heavy must the weight be, in order that, when it is removed, the box may float with only one-third of it immersed ?

(35) Find the specific gravity of a material such that a cylinder formed of it four inches long floats in water with three inches immersed.

(36) If a cubic foot of water weigh 1000 oz., and a cube whose edge is 18 inches weigh 2250 oz., how far will a cylinder whose length is 3 inches, formed of the same material as the cube, sink in water ?

(37) A body, whose specific gravity is 2·7 and weight in vacuo 3 lbs., when immersed in a fluid weighs 2 lbs.; find the specific gravity of the fluid.

(38) The specific gravity of mercury is 13·5 and that of aluminium is 2·6; how deep will a cubic inch of aluminium sink in a vessel of mercury ?

(39) If a body floats on a fluid two-thirds immersed, and it requires a pressure equivalent to 2 lbs. just to immerse it totally, what is the weight of the body ?

(40) If a body weighing 3 lbs. floats on a fluid one-half immersed, what pressure will sink it completely ?

(41) A piece of cork (s. g. = ·24) containing 2 cubic feet is kept below water by means of a string fastened to the bottom of a vessel ; find the tension of the string.

(42) Two bodies whose weights are w_1 and w_2 in air, weigh each w in water ; compare their specific gravities.

(43) The cavity in a conical rifle bullet is usually filled with a plug of some light wood. If the bullet be held in the hand beneath the surface of the water, and the plug be then removed, will the apparent weight of the bullet be increased or diminished ?

(44) A body, whose weight in air is 6 lbs., weighs 3 lbs. and 4 lbs. respectively in two different fluids ; compare the specific gravities of the fluids.

(45) A body whose specific gravity is 7·7 and weight in vacuo 7 lbs., when immersed in a fluid weighs 6 lbs. ; find the specific gravity of the fluid.

(46) A solid sphere floats in a fluid with three-fourths of its bulk above the surface : when another sphere half as large again is attached to the first by a string, the two spheres float at rest below the surface of the fluid ; shew that the specific gravity of one sphere is 6 times greater than that of the other.

(47) A piece of copper (s. g. = 8·85) weighs 887 grains in water, and 910 grains in alcohol; find the specific gravity of the alcohol.

(48) A uniform cylinder, when floating vertically in water, sinks a depth of 4 inches; to what depth will it sink in alcohol of specific gravity 0·79 ?

(49) A compound of silver (s. g. = 10·4) and aluminium (s. g. = 2·6) floats half immersed in a vessel of mercury (s. g. = 13·5). What weight of silver is there in 10 lbs. of the compound ?

(50) An iron rod weighing 10 lbs. is supported by means of a string, one-half of the rod being immersed in water. What force is exerted by the string, the specific gravity of iron being 7·8 ?

(51) A piece of silver weighing 1 oz. in air weighs ·905 oz. in water, what is its specific gravity ?

(52) Two bodies weighing in air 1 and 2 lbs. respectively are attached to a string passing over a smooth pulley; the bodies rest in equilibrium when they are completely immersed in water. If the specific gravity of the first body be twice that of water, find the specific gravity of the second.

(53) A cylinder 9 inches in height, specific gravity $\frac{4}{5}$, floats in water with its axis vertical; find the height of the surface of the cylinder above the surface of the water.

(54) Shew that if each division of the stem of the common hydrometer contains $\frac{1}{m}$th part of the bulk of the hydrometer, the ratio of the specific gravities of two fluids, in which the hydrometer floats with x and y divisions of the stem out of the fluid respectively, is equal to $m - y : m - x$.

(55) To a body which weighs 3 lbs. in air a piece of lead which weighs $5\frac{1}{2}$ lbs. in air is attached, and the two together weigh $1\frac{1}{4}$ lbs. in a fluid whose specific gravity is 4. Find the specific gravity of the body, that of lead being 11.

(56) A substance weighs 10 oz. in water and 15 oz. in alcohol, the specific gravity of which is ·7947 times that of water: find the number of cubic inches in the substance, taking the weight of a cubic foot of water as 1000 oz.

(57)   A block of ice, the volume of which is a cubic yard, is observed to float with $\frac{2}{25}$ths of its volume above the surface, and a small piece of granite is seen embedded in the ice ; find the size of the stone, the specific gravities of ice and granite being respectively ·918 and 2·65.

(58)   A cubical block of wood weighs 12 lbs. ; the same bulk of water weighs 320 oz. ; what part of the wood will be below the surface when it floats in water ?

(59)   A board 3 inches thick sinks 2¼ inches in water : what will a cubic foot of the same wood weigh, if a cubic foot of water weigh 1000 oz. ?

(60)   The specific gravity of beech-wood is ·85. What portion of a cubic foot of that wood will be immersed in sea water whose specific gravity is 1·03 ?

(61)   A cubical iceberg is 100 feet above the level of the sea, its sides being vertical.   Given the specific gravity of sea water = 1·0263 and of ice = ·9214, find the dimensions of the iceberg.

(62)   If a body of weight W float with three-quarters of its volume immersed in fluid, what will be the pressure on a hand which just keeps it totally immersed ?

(63)   Two hydrometers of the same size and shape float in two different fluids with equal portions above the surfaces ; and the weight of one hydrometer : that of the other :: $m : n$; compare the specific gravities of the fluids.

(64)   A hydrometer, loaded with 40 grains, sinks 4 inches lower when floating in a fluid whose specific gravity is ·3 than in water ; without the weight it rises in the water one-twelfth of an inch higher : find the weight of the hydrometer.

(65)   If the volume between two successive graduations on the stem of a hydrometer be $\frac{1}{1000}$th part of its whole bulk, and it floats in distilled water with 20 divisions, and in sea water with 46 divisions, above the surface ; find the specific gravity of sea water.

(66)   A piece of lead is found to weigh 13 lbs. in water, and when a block of wood weighing 6 lbs. is attached to it the two together weigh 9 lbs. in water.   Find the specific gravity of the wood.

(67) What is the weight of a hydrometer which sinks as deep in rectified spirits, specific gravity ·866, as it sinks in water when loaded with 67 grains ?

(68) The weight of a body A in water of specific gravity $=1$ is 10 oz., of another body B in air whose specific gravity $=·0013$ is 15 oz.; while A and B connected together weigh 11 oz. in water: shew that the specific gravity of B is 1·0713.

(69) A substance weighs 20 oz. in water and 25 oz. in alcohol, the specific gravity of which is ·7947 times that of water ; find the number of cubic inches in the substance, taking the weight of a cubic foot of water as 1000 oz.

CHAPTER V.

On the Properties of Air.

69. THE thin and transparent fluid which surrounds us on all sides, and which we call the Air or the Atmosphere, is a material body which possesses weight and resists compression. We can prove by experiment that even a small mass of air has an appreciable weight, by exhausting the air from a glass vessel (by a process which we shall describe in the next article). We then find that the vessel weighs less than it weighed before the air was taken out of it.

That the air resists compression is evident from the force required to drive down the piston of a syringe when the open end is closed.

Every body exposed to the atmosphere is subject to a pressure of nearly 15 pounds on each square inch of its surface. We feel no inconvenience from this great pressure, because the solid parts of our bodies are furnished with incompressible fluids, capable of supporting great pressures, while the hollow parts are filled with air like that which surrounds us. Also, since the atmosphere acts equally on all parts of our bodies, we have no difficulty in moving.

70. *Hawksbee's or the common Air Pump.*

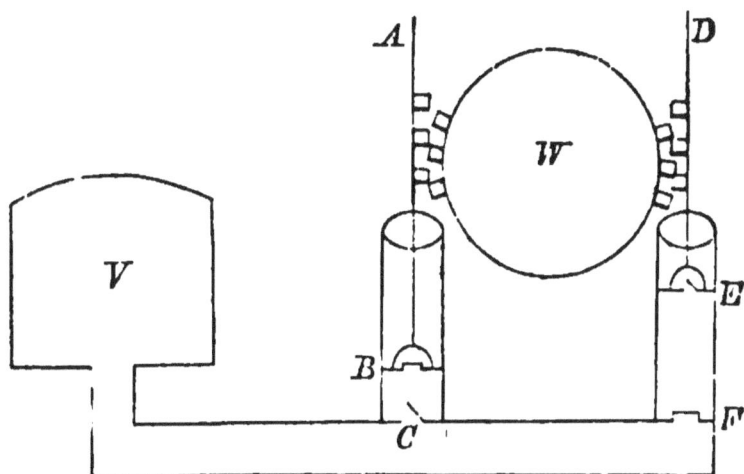

AB and *DE* are two pistons with valves opening upwards, which are worked up and down two cylindrical barrels by means of the toothed wheel *W* in such a way that one piston descends as the other ascends. The barrels communicate, by means of valves at *C* and *F* opening upwards, with a pipe leading into a strong glass vessel *V* called the receiver.

Suppose *B* to be at its lowest position and therefore *E* at its highest position. Then as *B* ascends the valve at *B* closes, and the air in the receiver and pipe opens *C* and expands itself in the barrel. As soon as *B* begins to ascend *E* begins to descend, the valve at *E* opens, the valve at *F* remains closed.

The air which before occupied the receiver and pipe, now occupies the receiver, the pipe, *and one of the barrels*, and is therefore rarefied.

Now let the wheel be turned back: then as *E* ascends the valve at *E* closes and *F* is opened, and meanwhile *B* is opened as it descends, and *C* being closed, a quantity of the rarefied air is taken from the receiver and pipe.

This process may be continued till the air in the receiver is so rarefied that it cannot lift the valves at *C* and *F*, and then the action of the instrument must cease.

71. *Smeaton's Air Pump.*

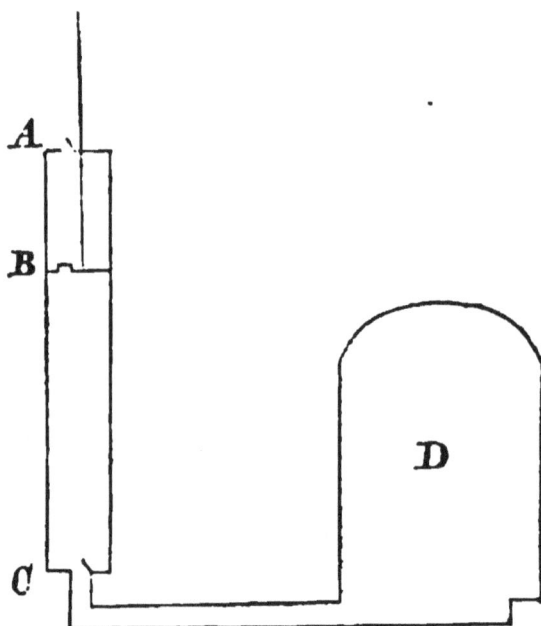

AC is a cylindrical barrel communicating with a strong vessel D called the receiver. At A and C, the ends of the barrel, are valves opening upwards.

A piston with a valve B opening upwards works up and down the barrel. Suppose the piston to be in its lowest position. Then as the piston ascends, the pressure of the air being removed from the upper surface of the valve at C, the air in DC opens C and expands into the barrel, while the valve at B is closed by the pressure of the atmosphere.

Thus a quantity of air is drawn away from the receiver. As soon as the piston begins to descend, the valve at A is closed, B opens and C is closed, and no external air comes into the barrel or receiver.

When the piston again ascends the air in the barrel is again drawn out.

The only limit to the exhaustion of the air by this pump arises from the difficulty in making the piston come into close contact with the valves at A and C.

NOTE. The advantage of Smeaton's Air Pump is that since the valve at A closes as soon as the piston begins to descend it relieves B from the pressure of the atmosphere, and the valve at B is opened by a very slight pressure from the air beneath. Hence this pump is capable of producing a greater degree of exhaustion than Hawksbee's.

72. To find the density of the air in the receiver of Smeaton's Air Pump after n ascents of the piston.

Let the measures of the capacities of the receiver and the barrel be respectively x and y.

Then the air which occupied the space whose measure is x when the piston was at C, will occupy the space whose measure is $x+y$ when the piston comes to A,

$$\therefore \frac{\text{density after one ascent}}{\text{density at first}} = \frac{x}{x+y},$$

$\therefore$ density after one ascent $= \dfrac{x}{x+y}$. (density at first).

Similarly,

density after second ascent $= \dfrac{x}{x+y}$. (density after one ascent)

$$= \left(\frac{x}{x+y}\right)^2. \text{(density at first)},$$

and so on ;

$\therefore$ density after nth ascent $= \left(\dfrac{x}{x+y}\right)^n$. (density at first).

The same formula is applicable to Hawksbee's Air Pump, if x represent the measure of the capacity of the receiver and pipe, and y the measure of the capacity of each of the barrels.

73. *The Barometer.*

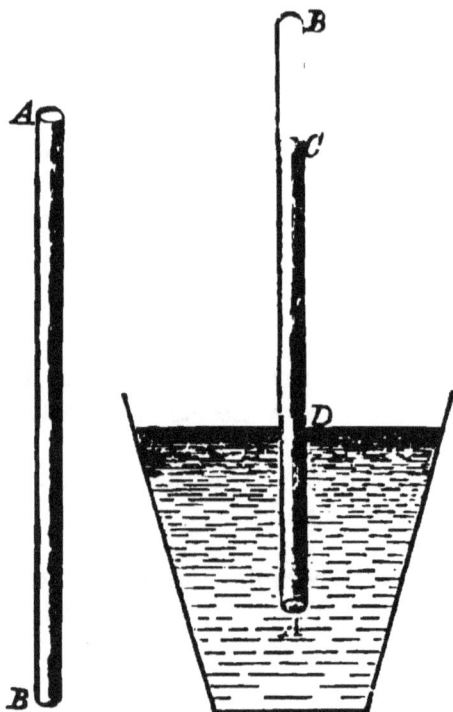

The Barometer is an instrument for measuring the pressure of the atmosphere.

If we take a glass tube about 32 inches long, open at *A* and closed at *B*, and fill it with mercury: if we then close the end *A*, invert the tube, place it in a vessel full of mercury, called the basin, and then remove the stoppage from *A*, the mercury in the tube will sink a little, leaving a vacuum in *BC*, and resting when the height of the column *CD*, that is, the distance of the surface of the mercury in the tube from the surface of the mercury in the basin, is from 28 to 31 inches.

That the column *CD* is supported by the pressure of the atmosphere may be shewn by placing the instrument under the receiver of an air pump. Then as the air is exhausted, the mercury will sink in the tube, and if all the air could be pumped out the mercury would sink entirely into the vessel. This experiment proves that the pressure of the air on the exposed surface of the mercury in the basin sustains the column of mercury in the tube.

74. *To shew that the pressure of the atmosphere is accurately represented by the weight of the column of mercury in the Barometer.*

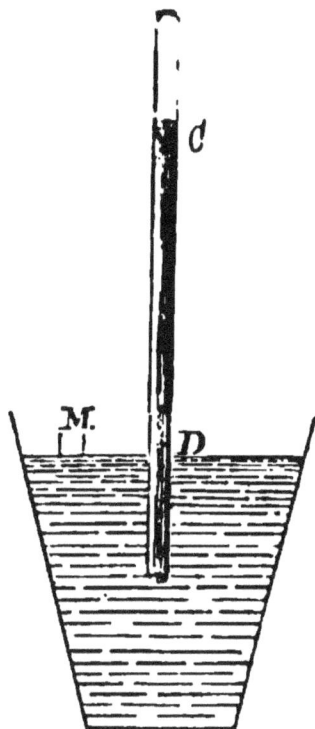

Take in the surface of the mercury in the basin an area M equal to the area of the horizontal section of the tube at D.

Then area M = area of the base of the column of mercury in the tube, and since these areas are equal and in the same horizontal plane, the pressures on them are equal.

Now pressure downwards on M = atmospheric pressure on area M, and pressure downwards at D = weight of column of mercury CD.

Therefore the atmospheric pressure on area M is equal to the weight of the column of mercury CD.

It follows then that the atmospheric pressure on any area is equal to the weight of the column of mercury in the barometer, having the same area for its base.

Consequently the weight of the column of mercury in the barometer is the proper representative of the pressure of the atmosphere on a given surface.

75. Hence it follows that the *height* of the column of mercury in the barometer is proportional to the pressure of the atmosphere.

If then we have a vertical tube of uniform bore filled up to the level *D* with mercury, if *D* be exposed to the atmospheric pressure and if *M* be some other level in the tube, and if *h* be the height of the barometric column,

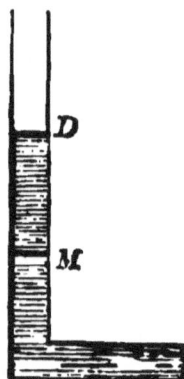

$$\frac{\text{pressure at } D}{\text{pressure at } M} = \frac{\text{weight of a column of mercury of height } h}{\text{weight of a col. of mercury of height } (h + DM)},$$

$$= \frac{h}{h + DM}.$$

76. *To find the Atmospheric Pressure on a Square Inch.*

The pressure of the atmosphere on a square inch is determined by finding the weight of a column of mercury whose base is a square inch and whose height is the same as the height of the column of mercury in the barometer.

Taking the specific gravity of mercury as 13·6, the weight of a cubic foot of distilled water as 1000 oz., and the height of the barometric column at the level of the sea as 30 inches, we have pressure of atmosphere on a square inch

$$= \left(30 \times 1 \times 1 \times \frac{1000}{1728} \times 13\cdot6\right) \text{ ounces,}$$

$$= \frac{30 \times 1000 \times 136}{1728 \times 10} \text{ ounces,}$$

$$= 236\tfrac{1}{9} \text{ ounces,}$$

$$= 14\tfrac{22}{144} \text{ lbs.}$$

77. In estimating the pressure at a point in the interior of fluid exposed to the atmospheric pressure, we must add to the pressure on a unit of area containing the point the atmospheric pressure on a unit of area.

Suppose for instance we have to find the pressure at a depth of 100 feet in a lake, (1) neglecting atmospheric pressure, (2) taking the atmospheric pressure into account.

Take a square inch as the unit of area: then

(1) Pressure at depth of 100 feet on a square inch

= weight of a column 'of water 100 feet in height, resting on a base of a square inch

= weight of a column of water whose cubic content is $(100 \times 12 \times 1 \times 1)$ cubic inches

$$= \left(\frac{1200}{1728} \times 1000 \right) \text{ oz.}$$

$$= \frac{1200 \times 1000}{1728 \times 16} \text{ lbs.}$$

$$= 43 \frac{29}{72} \text{ lbs.}$$

(2) Pressure at depth of 100 feet on a square inch

$$= \left(43 \frac{29}{72} + 15 \right) \text{ lbs. nearly,}$$

$$= 58 \frac{29}{72} \text{ lbs. nearly.}$$

78. The Atmosphere is most dense at the surface of the Earth, and its density diminishes with its height. Hence as one ascends a mountain the weight of the incumbent air is diminished, and the mercury in the barometer sinks. Thus the barometer furnishes a means of ascertaining approximately the height of a mountain.

79. A Barometer might be formed with any fluid, but mercury is preferred to other fluids because of its great density. A Water-barometer must have a tube of great length, since the atmosphere supports a column of water more than 13 times as high as the column of mercury supported in the mercurial barometer.

80. *The pressure of a given quantity of air, at a given temperature, varies inversely as the space it occupies.*

The following proof by experiment establishes the truth of this law.

ABC is a bent tube, cylindrical, uniform and vertical. The branch *AB* is much longer than the branch *BC*. The ends are open.

Mercury is poured drop by drop into the end *A* till the surface of the mercury in the two branches stands at the same level at *P* and *Q*.   The end *C* is then closed.

Then the pressure of air in *CQ*=the atmospheric pressure.

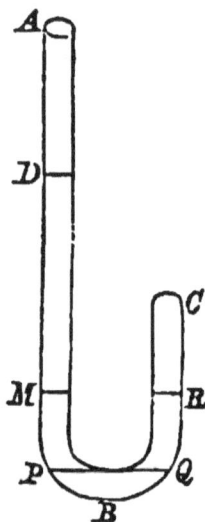

Let mercury be again poured in at *A*, (the effect of which is to compress the air in *CQ*,) till the surface of the mercury in the shorter branch stands at *R*, *halfway* between *C* and *Q*.

It is then found that the mercury in the longer branch will stand at a point *D*, such that the length of the column of mercury *DM* (*M* being level with *R*) is exactly equal to the height of the barometer at the time of making the experiment.

Now pressure at M = pressure at R.

But pressure at M = weight of column of mercury DM
+ pressure of atmosphere at D,

= atmospheric pressure + atmospheric
pressure = twice the atmospheric
pressure;

∴ pressure of the air in CR = twice the atmospheric pressure.

Hence the pressure of the air in CR is twice as great as
was the pressure of the air in CQ.

That is, when the given quantity of air in CQ has been
compressed into *half* the space, the pressure of the compressed
air is *twice* as great as it was at first.

81. The proof given in the preceding Article may be put
in a more general form, R being *any* point between C and Q,
thus :—

Let mercury be again poured in at A till the surface of
the mercury stands at D and R in the branches, and let M be
level with R.

Then it is found that if the spaces CQ, CR successively
occupied by the air be measured, and if h be the height of
the barometer at the time of performing the experiment,

$$\frac{\text{space } CQ}{\text{space } CR} = \frac{h + DM}{h}.$$

Now it is clear by Art. 75,

$$\frac{\text{pressure supporting air in } CQ}{\text{pressure supporting air in } CR} = \frac{h}{h + DM},$$

$$\therefore \frac{\text{pressure of air in } CQ}{\text{pressure of air in } CR} = \frac{CR}{CQ}.$$

Cor. Hence we can shew that the elastic force of air varies as its density.

For since the same quantity of air is confined in CQ and CR

density of air in CR : density of air in CQ

$$:: CQ : CR$$

$$:: \text{pressure of air in } CR : \text{pressure of air in } CQ.$$

82. *The Condenser.*

AC is a cylindrical barrel·with a valve at the bottom, C, opening downwards into a vessel B, called the receiver. A piston with a valve A, opening downwards, works in the barrel.

Suppose the piston to be at the top of the barrel. When the piston descends, the air in the barrel being condensed closes the valve at A, and opens the valve at C. Thus the air which was contained in the barrel is forced into the receiver. When the piston is raised again, the denser air in B keeps the valve at C closed, while the pressure of the atmosphere opens A, and the barrel is refilled with atmospheric air, which is forced into the receiver at the next descent of the piston.

The process may be continued till the required quantity of air has been forced into B.

83. To find the density of the air after n descents of the piston.

Let x and y bo the measures of capacities of the receiver and barrel respectively.

Then the air which occupied the space whose measure is $x + y$, when the piston was at the top of the barrel, will occupy the space whose measure is x when the piston comes to the bottom of the barrel;

$$\therefore \frac{\text{density of air in receiver after one descent}}{\text{density of air at first}} = \frac{x+y}{x},$$

$\therefore$ density of air after one descent $= \dfrac{x+y}{x}$. (density of air at first).

Similarly,

density after second descent $= \dfrac{x+2y}{x}$. (density of air at first)

and so on;

$\therefore$ density after nth descent $= \dfrac{x+ny}{x}$. (density of air at first).

EXAMPLES.—V.

(1) If the capacity of the receiver in Smeaton's Air Pump be ten times that of the barrel, what will be the exhaustion produced by six strokes of the piston?

(2) Find the pressure of the air in the receiver of an Air Pump after two strokes of the piston, the volume of the receiver being eight times that of the barrel.

(3) Find the ratio of the volume of the receiver to that of the barrel in the Air Pump, if at the end of the third stroke the density of the air in the receiver : the original density :: 729 : 1000.

(4) Is it necessary that the section of the tube through which the mercury rises in the barometer should be the same throughout ?

(5) Assuming that a cubic foot of water weighs 1000 oz. and a cubic inch of mercury weighs $7\frac{5}{8}$ oz., find the pressure on a square inch at a depth of 90 feet below the surface of the sea, when the barometer stands at 30 inches.

(6) If the area of a section of the basin of a barometer be 10 times that of a section of the tube, and the mercury fall $1\frac{1}{4}$ inches in the tube, find the true variation in the height of the mercury, and draw a figure representing the instrument.

(7) If a hole were made in the tube of a barometer, what would be the effect ?

(8) If the weight of the column of mercury which is above the exposed surface in a barometer be an ounce, and the area of the transverse section of the tube $\frac{1}{234}$ of a square inch, what is the pressure of the atmosphere on a square inch ?

(9) When the mercurial barometer stands at 30 inches, what will be the height of the column in a barometer filled with a fluid of specific gravity $3\cdot4$, the specific gravity of mercury being $13\cdot6$?

(10) If a piece of iron float in the mercury contained in the tube of a barometer, will it have any effect on the indication of the instrument ?

(11) If a body were floating on a fluid, with which the air was in contact, and the air were suddenly removed, would the body rise or sink in the fluid ?

(12) What would be the effect of admitting a little air into the upper part of the tube of the Barometer ?

(13) A pipe carries rain water from the top of a house to a large tank, the surplus water in which escapes through a valve in the top which rises freely. A weight of 21 lbs is placed on it, and it is found that the water rises in the pipe to the height of 20 feet before the valve opens. Find its area, assuming that the height of the Water-Barometer is 34 feet and the atmospheric pressure 15 lbs. on the square inch.

(14) A cylinder filled with atmospheric air, and closed by an air-tight piston, is sunk to the depth of 500 fathoms in the sea; required the compression of the air, assuming the specific gravity of sea-water to be 1·027, the specific gravity of mercury 13·57, and the height of the barometer 30 inches.

(15) A barometer is sunk to the depth of 20 feet in a lake: find the consequent rise in the mercurial column, the specific gravity of mercury being 13·57.

(16) If a body, exposed to the pressure of the air, float in water, prove that it will rise very slightly out of the water as the barometer rises, and sink a little deeper as the barometer falls.

(17) Water floats on mercury to the depth of 17 feet : compare the atmospheric pressure with the pressure at a point 15 inches below the surface of the mercury, taking into account the atmospheric pressure on the surface of the water, having given that the heights of the mercurial and water barometers are 30 inches and 34 feet respectively.

(18) Explain clearly why a balloon ascends.

(19) Explain how it is that a bladder filled with air, will, if conveyed deep enough in the sea, sink to the bottom.

(20) What would be the height of the column of mercury (s. g. = 13·56) corresponding to a pressure of 14 lbs. 2 oz. on the square inch ?

(21) A cubical vessel full of air, whose edge equals 6 inches, is closed by a weightless piston.   Find the number of pounds which must be placed on the piston in order that it may rest in equilibrium at a distance of 2 inches from the bottom of the vessel : the pressure of the atmosphere being 15 lbs. on a square inch.

(22) The lower valve of a pump is 30 feet 4 inches above the surface of the water to be raised : find the height of the barometer when the pump ceases to work, the specific gravity of mercury being 13·6.

(23) It is found that the cork of a bottle is just driven out
when the pressure of the air within is double that without ; the
bottle is then filled with mercury and inverted, and it is again
found that the cork is just driven out.   Given that the
barometer was standing at 30 inches at the time, find the
height of the bottle.

(24) Find the ratio of the volume of the receiver to that
of the barrel in a Condenser, if at the end of the third stroke
the density of the air in the receiver : its original density
:: 3 : 2.

(25) A hollow cylinder closed at the upper end and open
at the lower is depressed from the atmosphere into water, its
axis being kept vertical, and is found to float with its upper
end in the surface of the water.   What will be the effect on
the cylinder of an increase of atmospheric pressure ?

(26) If the volume of the cylinder in a Condenser be one-
fifth the volume of the receiver, find the pressure at any
point of the latter after 20 strokes.

(27) The pressure at the bottom of a well is double that
at the depth of a foot; what is the depth of the well if the
pressure of the atmosphere be equivalent to 30 feet of water ?

(28) A cubic foot of water weighs 1000 oz. ; what will be
the pressure on each square inch of the base of a cube whose
edges are 10 inches, when filled with water ?

(29) A cubic foot of water weighs 1000 ounces, and the
pressure of the air on a square inch is 236 ounces ; find the
pressure on 16 square inches at a depth of 9 feet below the
surface of a pond.

(30) If A, B, C, be three points in a uniform fluid at rest,
the three points being in the same vertical line, and the difference of the pressures at A and B : difference of the pressures at A and C as $p : q$, find the ratio of AB to BC.

(31) Explain the principle of the Air-gun.

(32) If the area of the basin of a barometer be 17 times that of a section of the tube, how ought the stem to be graduated in order that the reading may give the true height of the barometer?

(33) If the specific gravity of mercury be 13·57, and the weight of a cubic inch of water 252·6 grains, find the pressure of the air on a square inch in lbs., when the mercury in the barometer stands at 30·5 inches.

(34) If the tube of a barometer be 36 inches long, and, on account of air being in the upper part, the instrument stands at 27 inches, when a correct instrument stands at 30 inches, what length of tube would the air fill when reduced to atmospheric density?

(35) The specific gravity of the weights employed by jewellers, for weighing precious stones, is greater than that of the stones themselves. Is it more advantageous for the jeweller to sell stones when the barometer is high, or when it is low?

(36) A tube closed at both ends and 28 inches long is half filled with mercury, the remaining portion being occupied with air at atmospheric pressure. If the tube be placed in a vertical position with the mercury uppermost, and the upper end be opened, find how far the mercury will sink, the height of the barometer at the time being 28 inches.

CHAPTER VI.

On the Application of Air.

84. *The Diving Bell.*

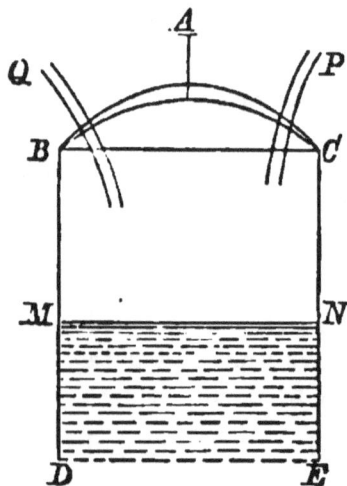

If a glass be inverted, and with its mouth horizontal be pressed down into a basin of water, it will be seen that though some portion of water ascends into the glass, the greater part of the glass is without water.

This is caused by the compression of the air, which prevents the water from rising in the glass.

The Diving Bell works on the same principle. A heavy iron chest *BCED*, open at *DE*, is suspended from a rope *A*, and lowered into the water, with its open end downwards. The water will then rise till the air in the chest is sufficiently compressed to prevent the water from rising beyond a certain height *MN*.

Air is pumped in occasionally through a pipe *P*, and the impure air is allowed to escape through another pipe *Q*.

85. *The Common or Suction Pump.*

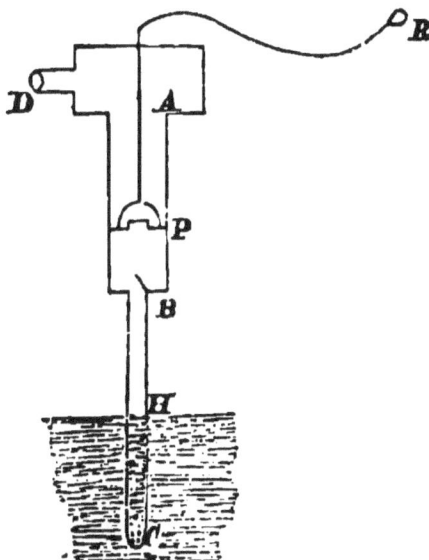

AB is a cylindrical barrel in which a piston *P*, with a valve opening upwards, is worked up and down by the handle *R*. *BC* is a pipe, communicating with the barrel by a valve, opening upwards. The end *C*, which is pierced with a number of small holes, is placed under the surface of the water which is to be raised.

Suppose the piston to be at the bottom of the barrel. Then when the piston is raised the valve *P* is closed by the pressure of the air on its upper surface, and there being little or no air in *PB*, the valve *B* is opened by the action of the air in *BC*, and as it continues open during the whole ascent of the piston, the air in *BH*, the part of the suction-pipe above the surface of the water, expands into the barrel, and becomes less dense than the air which presses on the water outside the suction-pipe. The water is consequently forced up the pipe by the pressure of the atmosphere, till the pressure downwards at *H* is equal to the atmospheric pressure.

When the piston descends the valve *B* closes, and the air in *PB*, being condensed, opens the valve *P*.

This process being continued, the water will at length rise through the valve *B*, and at the next ascent of the piston a mass of water will be lifted and discharged through the spout *D*.

The distance *BH* must be less than the height of a column of water which the atmospheric pressure can sustain, that is, less than 32 feet.

86. *The Forcing Pump.*

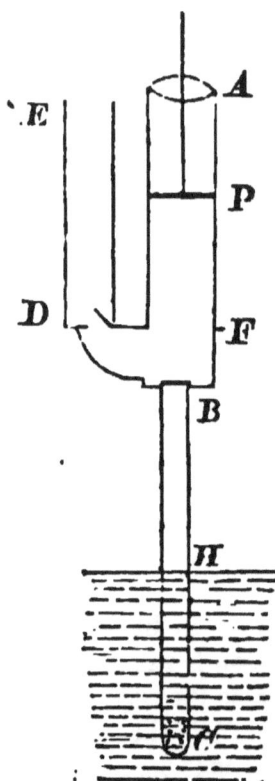

AB is a cylindrical barrel in which a solid piston *P* is worked up and down the space *AF*.

BC is a suction-pipe of which the end *C* is placed under the surface of the water.

DE is a pipe communicating with the barrel.

At *B* and *D* are valves opening upwards.

Suppose the piston to be at the bottom of its range in the barrel. Then when the piston is raised the valve at *D* remains closed, the air in *DBF* expands as the piston rises, and the air in *BH* opens the valve *B* and expands into the barrel. The water is therefore forced up the suction-pipe by the pressure of the atmosphere.

When the piston descends the air in *PFBD* is condensed, closes the valve *B*, opens the valve *D*, and escapes through *D*.

When the piston ascends again the water rises higher in *BC*, and this process is continued till the water rises through *B*. Then the piston on its descent forces the water up the pipe *DE*.

87. In order to produce a *continuous* stream through the pipe at *E*, the pipe is introduced into an air-tight vessel *DH* into which the valve *D* opens.

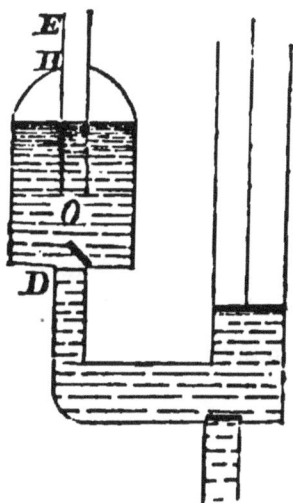

When the water has been forced into this vessel till it rises above *O*, the lower end of the pipe, the air which lies between the surface of the water in the vessel and the top of the vessel is suddenly condensed at each stroke of the piston, and by its reaction on the water forces it through the pipe *OE* in a continuous stream.

83. *The Fire Engine.*

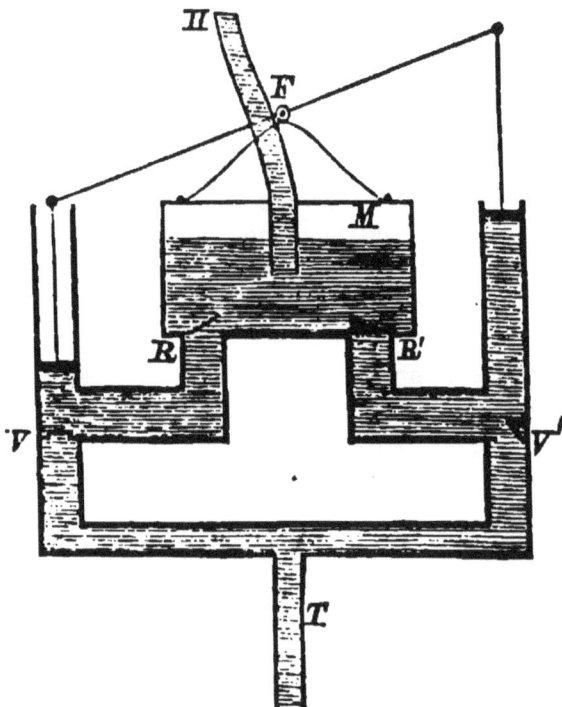

This machine consists of a double forcing-pump, both pumps communicating with the same air-vessel M.

The pipe T descends into a reservoir of water.

The valves opening upwards are at V, V' and R, R'.

F is a fixed beam round which the piston-rods work.

The water is discharged through the pipe H.

89. *The Lifting Pump.*

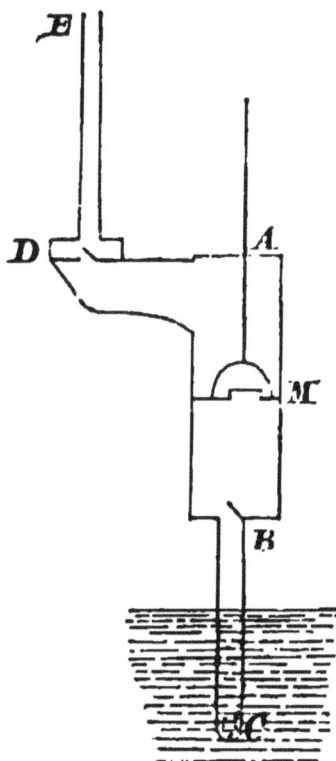

AB is a cylindrical barrel in which a piston with a valve *M* opening upwards works, the piston rod passing through an airtight collar at *A*.

BC is the suction-pipe of which the end *C* is placed under the surface of the water.

DE is a pipe up which the water is to be raised.

At *D* and *B* are valves opening upwards.

The water will be brought within reach of the piston by a process similar to that which has been described in the case of the other pumps.

When the piston ascends lifting water the valve at *D* opens, and the water is discharged into the pipe *DE*. When the piston descends, the valve at *D* closes, and prevents the return of the water in *DE* into the barrel.

Each stroke of the piston increases the quantity of water in *DE*, and thus the water may be raised to any height, provided that the barrel *AB*, the pipe *ED*, and the piston rod be strong enough to bear the pressure of the superincumbent column of water.

90. *The Siphon.*

The Siphon is a bent hollow tube of uniform bore, having one branch longer than the other. The tube is filled with fluid, the ends are closed, and the shorter branch is placed in a vessel containing fluid like that with which the siphon is filled.

Let the plane of the fluid's surface meet the branches of the siphon in H, K.

Then if the ends A, C be opened at the same moment, the

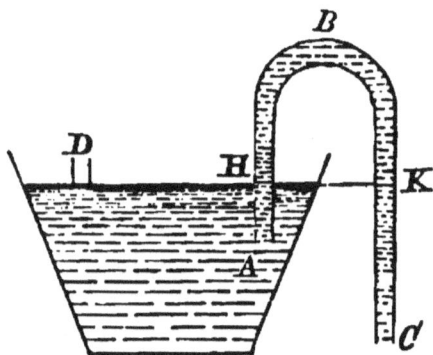

fluid will flow from C in a continuous stream till the vessel is emptied down to the level of A, provided that B, the highest point of the siphon, is at a less distance above the surface of the fluid than the height of a column of the fluid which the pressure of the atmosphere will sustain.

To explain this, consider the pressure on an area D, equal to the area of a horizontal section of the siphon, in the surface of the fluid : then

pressure of atmosphere at H in direction $HB =$ pressure on area D,

pressure of atmosphere at C in direction $CB =$ pressure on area D,

$\therefore$ pressure of atmosphere at H in direction $HB =$ pressure of atmosphere at C in direction CB.

Now pressure of atmosphere at H is diminished by the weight of column of fluid BH, and pressure of atmosphere at C is diminished by the weight of column of fluid BC, and since tho column *BC* is greater than column *BH*, the *effective* pressure of atmosphere in direction *HB* is greater than the *effective* pressure of atmosphere in direction *CB*, and therefore the fluid will be driven by the effective atmospheric pressure in a continuous stream in the direction *HBC*.

91. *On intermitting Springs.*

Intermitting Springs are springs which run for a time, then stop for a time, and then begin to run again.

This phenomenon is explained by the principle of tho Siphon.

Let *A* be a reservoir in a hill in which water is gradually collected through fissures, as *B*, *C*, *D*, communicating with tho external air.

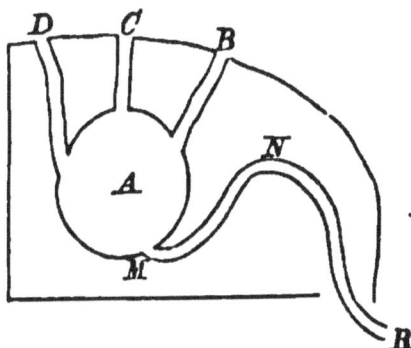

Now suppose a channel *MNR* to run from *A*, first ascending to *N* and then descending to *R*, a place lower than the reservoir.

As the water collects in *A* it gradually rises in the channel to *N*, and then flows along *NR*, and by the principle of the Siphon it will continue to flow till *A* is completely drained. Then the flow ceases till the water in *A* has collected sufficiently to reach *N*.

92. *Bramah's Press.*

The Hydrostatic Press, generally called Bramah's Press, is a machine by which an enormous pressure is obtained by means of water, the only assignable limits to its power being the strength of the materials of which it is formed.

AC is a forcing-pump, by the action of which water is forced into a tube *BD*, which has a valve *B* opening inwards.

E is a strong cylindrical piston, with a base many times larger than the base of the piston *A*, working in a water-tight collar at *M, N.*

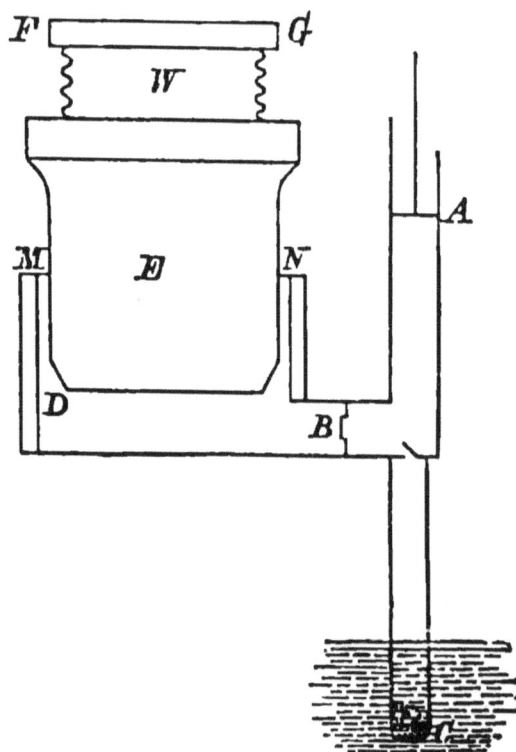

Between the top of the piston *E* and a fixed beam *FG*, a bale of goods, such as paper, cotton or wool, is placed.

Suppose the area of the base of *E* to be 200 times that of the base of *A*.

Then if a pressure of 100 lbs. be applied to *A*, a pressure of (200 × 100) lbs. or 20,000 lbs. will be conveyed to the base of *E*.

Thus any amount of pressure may be applied to *W*, either by increasing the pressure applied to *A*, or by making the base of *E* larger in comparison with the base of *A*.

EXAMPLES.—VI.

(1) What will be the effect of making a small aperture in the barrel of a Forcing Pump? If the piston work uniformly up and down the length of the barrel, and a small aperture be made one-third of the way up the barrel, how much more time than before will be consumed in filling a tank?

(2) If the upward motion of the piston of a Common Pump be stopped, when the water has risen to the height of 16 feet in the supply pipe, but has not yet reached the piston, find the tension of the piston-rod, the area of the piston being 4 square inches, and the atmospheric pressure 15 lbs. on the square-inch.

(3) What would be the effect of opening a small hole at any point in the Siphon, first above, secondly below the surface of the fluid in the vessel?

(4) What is the greatest height above the surface of a spring over which its water may be carried by means of a siphon-tube, when the barometer stands at 29 inches, the specific gravity of mercury being 13·57?

(5) What would take place in a siphon at work if the pressure of the atmosphere were removed?

(6) Will the siphon act better at the top or the bottom of a mountain?

(7) Could a siphon be employed to pump water out of the hold of a ship floating in a harbour?

(8) What is the greatest height over which water can be carried by means of a siphon when the mercurial barometer stands at 30 inches?

(9) If the ends of a siphon were immersed in two fluids of the same kind and the air were removed, describe what would take place.

(10) A hollow tube is introduced into the bottom of a cylindrical vessel through an air-tight collar; and a large tube, of which the top is closed, suspended over it, so as not quite to touch the bottom: consider the effect of gradually pouring water into the cylinder, until it reaches the level of the top of the inverted tube.

(11) A siphon is placed with one end in a vessel full of water, and the other in a similar empty one, both of which are on the plate of an air-pump. As soon as the water has covered the lower end of the siphon, a receiver is put on, and the air rapidly exhausted, and then gradually readmitted : describe the effects produced.

(12) A siphon, filled with water, has its ends inserted in vessels filled with water ; state what will take place when the vertical distances of the highest point of the siphon above the surface of the fluid are both less, both greater, and one greater and the other less than the height of the Water-Barometer.

(13) What is the length of the smallest siphon that can empty a vessel 2 feet deep ?

CHAPTER VII.

On the Thermometer.

93. THE general consequence of imparting heat to bodies is the expansion of their volume.

The particles which compose a *solid* body, as for instance a block of lead, are held together by the force of cohesion. It requires a force of great magnitude to increase or to decrease the volume of a block of lead, though lead is a soft metal. The application of heat, by weakening the force of cohesion, reduces lead and other metals to a liquid state, pushes the particles more widely apart, and thus increases the volume of the bodies to which it is applied.

If heat be applied to a *liquid*, as water, the cohesion of the particles is weakened, and they ultimately acquire a tendency to break away from each other and assume the form of a vapour.

If heat be applied to an *elastic fluid*, as air, it causes it to expand. Thus if a bladder, partly full of air, be placed before a fire, the air will expand and distend the bladder.

Again, if a piston *P* exactly fits a cylindrical tube *AB*, and is supported by the condensed air in *PB*, if heat be applied to the air in *PB* it will expand and raise the piston.

94. The Mercurial Thermometer is an instrument constructed to measure temperatures by means of the extent of the expansion or contraction of mercury.

It consists of a glass tube of uniform bore closed at *A* and terminating at the other end in a bulb. The bulb contains mercury, which extends part of the way up the tube. The space between the mercury and the top of the tube is a vacuum.

If the mercury in the instrument be subjected to an increase of heat, it expands and rises higher in the tube.

A vacuum is obtained in the upper part of the tube before the end *A* is closed by making the mercury in the instrument boil, so as to expel the air through the opening at *A*, which is then hermetically sealed, and the mercury sinking as it cools leaves a vacuum in the upper part of the tube.

95. *To graduate a Thermometer.*

The portion of the instrument containing the mercury is plunged into melting ice: the mercury shrinks, the column descends and finally becomes stationary. The point at which it rests is marked: it is the *freezing point* of the thermometer.

The instrument is next placed in the vapour of water boiling under a given atmospheric pressure: the mercury expands, the column rises and finally becomes stationary. The point at which it rests is marked : it is the *boiling point* of the thermometer.

The space between the freezing point and the boiling point is divided into equal spaces, called degrees.

In the Centigrade Thermometer freezing point is marked 0^{0} and boiling point 100^{0}.

In Fahrenheit's Thermometer freezing point is marked 32^{0} and boiling point 212^{0}.

In Reaumur's Thermometer freezing point is marked 0^{0} and boiling point 80^{0}.

96. *Having given the number of degrees on Fahrenheit's Thermometer, to find the corresponding number of degrees on the Centigrade Thermometer.*

Let AM be the line at which the mercury stands at freezing point,

BN .... at boiling point.

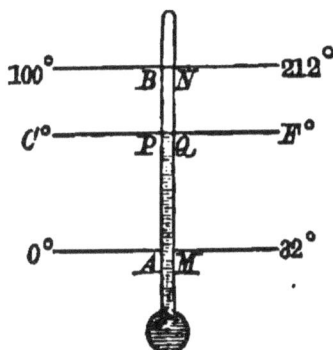

Then

AM and BN are marked 0° and 100° on the Centigrade scale
.... 32° and 212° .... Fahrenheit ......

Let the mercury stand at the line PQ, and suppose the graduations on the scales to be C° and F° respectively.

Now $\dfrac{AP}{AB} = \dfrac{MQ}{MN}$,

or $\dfrac{C}{100} = \dfrac{F-32}{212-32}$,

or $\dfrac{C}{100} = \dfrac{F-32}{180}$;

$\therefore \dfrac{C}{5} = \dfrac{F-32}{9}$,

and from this equation we can find C when F is given and F when C is given.

97. To compare the scales of the Centigrade and Reaumur's Thermometer, we proceed in the same way, putting 0°, R, 80° instead of 32°, F, 212° respectively, and we obtain

$$\frac{C}{100} = \frac{R}{80},$$

or $\dfrac{C}{5} = \dfrac{R}{4}.$

Hence the three scales are thus connected,

$$\frac{C}{5} = \frac{F-32}{9} = \frac{R}{4}.$$

98. The following examples will shew how to find the number of degrees marked on any one of the three scales when the number marked on one of the other scales is given.

Ex. (1) What reading on the Centigrade scale corresponds to $56°$ Fahrenheit ?

$$\text{Since } \frac{C}{5} = \frac{F-32}{9},$$

and $F = 56$,

$$\frac{C}{5} = \frac{56-32}{9};$$

$$\therefore 9C = 5 \times 24,$$

$$\therefore C = \frac{120}{9} = 13\tfrac{1}{3},$$

$\therefore$ the reading on the Centigrade scale is $13\tfrac{1}{3}$ degrees.

Ex. (2) What reading on the Fahrenheit scale corresponds to $14°$ Centigrade ?

$$\text{Since } C = 14,$$

$$\frac{14}{5} = \frac{F-32}{9};$$

$$\therefore 126 = 5F - 160,$$

$$\therefore 5F = 286,$$

$$\therefore F = 57\tfrac{1}{5},$$

that is, the reading on the Fahrenheit scale is $57\tfrac{1}{5}°$.

Ex. (3) If the sum of the readings on a Centigrade and a Reaumur be 90, what is the reading on each ?

Here we have two equations, from which we can find C and R,

$$\frac{C}{5} = \frac{R}{4} \ \ldots\ldots (1),$$

$$C + R = 90 \ \ldots\ldots(2);$$

$$\therefore \left. \begin{array}{l} 4C = 5R \\ 4C + 4R = 360 \end{array} \right\};$$

$$\therefore 4R = 360 - 5R,$$

$$\therefore 9R = 360,$$

and so $R = 40$ and $C = 50$.

EXAMPLES.—VII.

(1) Give the number of degrees in the Centigrade and Reaumur's scale respectively that correspond to the following readings on Fahrenheit's scale,

(1) 30°, (2) 45°, (3) 56°, (4) 0°, (5) -7°, (6) -45°.

(2) Give the number of degrees in the Centigrade and Fahrenheit's scale respectively that correspond to the following readings on Reaumur's scale,

(1) 5°, (2) 20°, (3) 0°, (4) -18°, (5) -64°, (6) 120°.

(3) Give the number of degrees in Fahrenheit's and Reaumur's scales respectively that correspond to the following readings on the Centigrade scale,

(1) 16°, (2) 45°, (3) 110°, (4) 0°, (5) -15°, (6) -24°.

(4) Is it necessary that the section of the tube through which the mercury rises in the Thermometer should be the same throughout ?

(5) If the sum of the readings on a Centigrade and Fahrenheit be 60, what is the reading on each ?

(6) At what temperature will the degrees on Fahrenheit be five times as great as the corresponding degrees on the Centigrade ?

(7) At what point do Fahrenheit and the Centigrade mark the same number of degrees ?

(8) Show how to graduate a Thermometer on whose scale 20° shall denote the freezing point, and whose 80th degree shall indicate the same temperature as 80° Fahrenheit.

(9) What will be the reading on the Centigrade when Fahrenheit stands at 78° ?

(10) The sum of the number of degrees indicating the same temperature on the Centigrade and Fahrenheit is 88, find the number of degrees on each.

(11) What reading on the Centigrade corresponds to 49° Fahrenheit ?

(12)   What would be the inconvenience of having the bore of the Thermometer large ?

(13)   At what temperature will the degrees on Fahrenheit be 3 times as great as the corresponding degrees Centigrade ?

(14)   The numbers of degrees indicated at the same instant by a Centigrade and a Fahrenheit's thermometer are as 5 : 17; determine the temperature.

(15)   What is the temperature when the number of degrees on the Centigrade is as much below zero, as that on Fahrenheit's is above zero ?

(16)   One Thermometer marks two temperatures by 9° and 10° ; another Thermometer by 12° and 14°; what will the latter mark, when the former marks 15° ?

(17)   One Thermometer marks two temperatures by 8° and 10° ; another Thermometer by 11° and 14°; what will the latter mark when the former marks 16°?

(18)   If the difference of the readings on Fahrenheit and Reaumur be 47, what are the readings ? If the difference increase by a given number of degrees, find how much each of the thermometers has risen.

CHAPTER VIII.

Miscellaneous Examples.

99. WE shall now give a series of examples to illustrate more fully the principles explained in the preceding Chapters. The important law of pressure in the case of compressed air, of which we treated in Arts. 80, 81, will be referred to as *Marriotte's Law* *.

Examples worked out.

1. *Water is 770 times as heavy as air. At what depth in a lake would a bubble of air be compressed to the density of water, supposing Marriotte's law to hold good throughout for compression?*

At the surface the density = that of atmosphere,

and 33 feet of water are equivalent to one atmosphere ;

∴ at depth of 33 ft. the density = twice atmospheric pressure,

.... (2 × 33) ft. .... = three times ....

.... (769 × 33) ft. .... = 770 times ....

∴ the density will be equal to that of water at a depth of

(769 × 33) ft. *i. e.*, 25377 ft.

---

* It was proved by the independent researches of Marriotte, a French Physician, and Boyle, the English Philosopher.

2. *A body weighs in air 1000 grs., in water 300 grs., and in another liquid 420 grs.: what is the specific gravity of the latter liquid?*

In water the body loses (1000 − 300) grs., *i. e.* 700 grs.,

in other liquid ....(1000 − 420) grs., *i. e.* 580 grs.;

∴ equal volumes of water and of the other liquid weigh respectively 700 grs. and 580 grs.

∴ measure of specific gravity of other liquid $= \dfrac{580}{700} = \cdot 8285714$.

3. *Taking account of atmospheric pressure, and taking 33 feet as the height of the water barometer, at what depth in a lake is the pressure twice what it is at a depth of one yard?*

Pressure at the surface = weight of column of water 33ft. high,

Pressure at 3ft. depth = weight of column of water 36ft. high;

∴ for a double pressure we must take 36 feet lower, that is, 36 feet lower than 3 feet, or 39 feet from the surface.

4. *A flat piece of iron, weighing 3 lbs., floats in mercury; and if another piece of iron of like density weighing* $2\dfrac{5}{26}$ *lbs. is placed upon it, the flat piece is just immersed. Compare the specific gravities of iron and mercury.*

Total weight of iron $= \left(3 + 2\dfrac{5}{26} \right)$ lbs. $= 5\dfrac{5}{26}$ lbs.

The volumes of the part immersed and of the whole will be as the weights, that is, as $3 : 5\dfrac{5}{26}$, or as 78 : 135.

∴ sp. gr. of iron : sp. gr. of mercury = 78 : 135,

$= 26 : 45$.

5. *Air is confined in a cylinder surmounted by a piston without weight, whose area is a square foot. What weight must be placed on the piston that the volume of air may be reduced to half its dimensions?*

By Marriotte's law the air when reduced to half its volume will have double its original pressure. Hence taking 15 lbs. per square inch as the original atmospheric pressure, it becomes 30 lbs. per square inch below the piston. But the atmosphere still exerts a pressure of 15 lbs. per square inch above the piston. Therefore a pressure of 15 lbs. more per square inch is required to keep the piston at rest.

$$\therefore \text{ weight required} = (15 \times 144)\,\text{lbs.} = 2160\text{ lbs.}$$

6. *If the capacity of the receiver of an air-pump be* 10 *times that of the barrel, shew that, after* 3 *strokes of the piston, the air in the receiver will have lost nearly one-fourth of its density.*

By the formula of Art. 72, if ρ_0 and ρ_n be the densities originally and after the nth stroke, and R and B be the capacities of the receiver and barrel,

$$\frac{\rho_n}{\rho_0} = \left(\frac{R}{R+B}\right)^n,$$

$$\therefore \frac{\rho_3}{\rho_0} = \left(\frac{10}{10+1}\right)^3 = \frac{1000}{1331};$$

$$\therefore \text{density lost} = \left(1 - \frac{1000}{1331}\right)\rho_0 = \frac{331}{1331}\rho_0 = \frac{1}{4}\rho_0 \text{ nearly.}$$

7. *A block of wood* $\left(\text{ s. g. } \frac{12}{13}\right)$ *weighing* 156 *lbs. is floating in fresh water. What weight placed on it will sink it to the level of the water?*

Let $x =$ the weight in lbs.

Then $x + 156 =$ weight in lbs. of water displaced by volume of wood alone,

$$= \frac{13}{12} \times 156,$$

$$= 169;$$

$$\therefore x = (169 - 156)\,\text{lbs.} = 13\text{ lbs.}$$

8. *In a mixture of two fluids, of which the specific gravities are* 3 *and* 5 *respectively, a body, whose s. g. is* 8, *loses half its weight. Compare the volumes mixed.*

Weight lost = weight of fluid displaced,

$$= \frac{1}{2} \text{ weight of body whose s. g. is 8,}$$

$\therefore$ s. g. of the mixture is 4.

And since the separate specific gravities are 3 and 5, while the sp. gr. of the mixture is 4, the fluids must be mixed in equal volumes.

9. *A vessel of water has for its horizontal section a rectangle 6 feet by 2 feet. A substance weighing 550 lbs. is immersed in it, and the water rises 8 inches. Find the specific gravity of the substance.*

Sectional area $= 12$ square feet.

Volume of substance $= \left(12 \times \dfrac{2}{3} \right)$ cub. ft.

$$= 8 \text{ cubic feet;}$$

$\therefore$ 8 cubic feet of the substance weigh 550 lbs.;

$\therefore$ 1 cub. ft. .... $\dfrac{550}{8}$ lbs., or 68·75 lbs.

Also, a cubic foot of water weighs 62·5 lbs.,

$$\therefore \text{ sp. gr. of substance} = \frac{68\cdot75}{62\cdot5} = 1\cdot1.$$

10. *A cylinder floats in a fluid A with one-third of its axis immersed, and in another B with three-fourths of its axis immersed. How deep will it float in a fluid which is a mixture of equal volumes of A and B?*

Sp. gr. of A : sp. gr. of $B = \dfrac{3}{4} : \dfrac{1}{3}$,

$$= 9 : 4;$$

$\therefore$ sp. gr. of mixture of equal volumes $= \dfrac{9+4}{2} = 6\cdot5.$

If therefore the body has $\dfrac{1}{3}$ of its axis immersed in a fluid of s. g. 9, when it is immersed in a fluid of s. g. 6·5 the part immersed is obtained from the following relation, where x is the part immersed,

$$6\frac{1}{2} : 9 = \frac{1}{3} : x,$$

$$\therefore x = \frac{9 \times \dfrac{1}{3}}{6\dfrac{1}{2}} = \frac{6}{13}.$$

100. We shall now give a set of easy Examples to be worked by the student by way of practice.

Examples.—VIII.

1. An iceberg (s. g. ·925) floats in sea-water (s. g. 1·025). Find the ratio of the part out of the water to the part immersed.

2. A body floats in a fluid (s. g. ·9) with as much of its volume out of the fluid as would be immersed if it floated in a fluid (s. g. 1·1). Find the specific gravity of the body.

3. Find the Fahrenheit Temperatures corresponding to $-40°$ and $+350°$ Centigrade.

4. The capacities of the barrel and receiver in a Smeaton's air-pump are as 1 : 3. A barometer enclosed in the receiver stands at 28 inches. What will be the height after three upward strokes of the piston ?

5. Two hydrometers of the same size and shape float in two different fluids with equal portions above the surfaces, and the weight of one hydrometer : that of the other $= 1 : p$. Compare the specific gravities of the fluids.

6. A man weighing 10 stone 10 oz. floats with the water up to his chin when he has a bladder under each arm equal in size to his head and without weight. If his head be one-twelfth of his whole bulk, find his specific gravity.

7. At what height does the water barometer stand when the mercurial barometer stands at 28 inches (s. g. of mercury $= 13·6$)?

8. What degree Centigrade corresponds to $27°$ Fahrenheit ?

9. A man 6 feet high dives vertically downwards with his hands stretched 18 inches beyond his head. What depth has he reached when the pressure at his fingers' ends is $\frac{3}{2}$ that at his feet ?

10. A string will bear a strain of 10 lbs. 7 oz. Determine the size of the largest piece of cork (s. g. ·24) which it can keep below the surface of mercury (s. g. 13·6).

11. In De Lisle's Thermometer the freezing point is 150° and the boiling point zero. What degree of this thermometer corresponds to 47° Fahrenheit?

12. Cork would float in n atmospheres. Find n (s. g. of air and cork being ·0013 and ·24).

13. An elastic body of s. g. ·5 is compressed to $\dfrac{20+n}{20+4n}$ of its natural size by immersion n feet in water. At what depth will it rest?

14. If the body in Question 13 weigh 10 lbs., what are the magnitudes and directions of the forces which will keep it in equilibrium at depths (a) 5 feet, and (β) 30 feet?

15. At what depths will the force required to keep the body in Questions 13 and 14 at rest be 1 lb.?

16. At what temperature are the readings on Reaumur, Centigrade and Fahrenheit proportional to 4, 5, 25?

17. At what temperature is the sum of the readings on Reaumur, Centigrade and Fahrenheit 212?

18. A body (s. g. 2·6) weighs 22 lbs. in vacuo and another body (s. g. 7·8) weighs n lbs. in vacuo; and their apparent weights in water are equal. Find n.

19. Find the specific gravity of the fluid in which the apparent weights of 1 lb. of one substance (s. g. 3) and 3 lbs. of another substance (s. g. 2·25) are equal.

20. Equal volumes of two substances (s. g. 2·7 and 6·1) are immersed in water and balance on a straight lever 71 inches long. Find the position of the fulcrum.

101. We proceed with some examples of somewhat greater difficulty than those already given.

Note. We shall assume that the volume of a sphere is $\frac{4}{3}\pi r^3$, r being the radius.

Examples worked out.

1. *Show how the depth of the descent in a Diving Bell can be determined from observations on the barometer.*

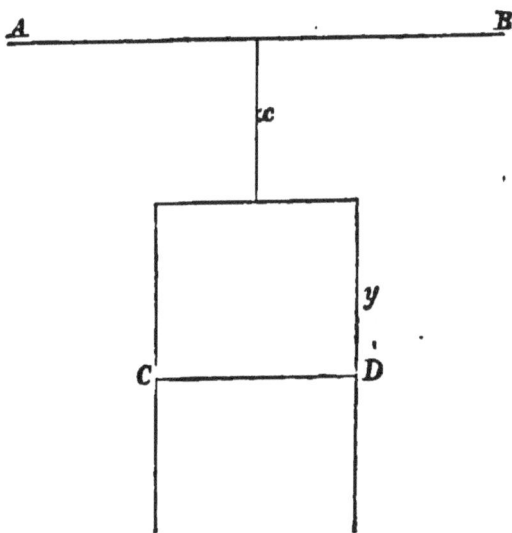

Let AB be the surface of the water, CD the water level in the bell at the end of the descent.

Now pressure at CD is equal to pressure throughout the upper part of the bell, and is therefore equal to the pressure due to atmosphere + weight of column of water $(x+y)$ ft. high.

Hence if S be the measure of the specific gravity of mercury, and h, h' the measures of the heights of the mercurial column at surface of the water and at the bottom,

measure of pressure at $CD = hs + (x+y) \times 1$.

But measure of pressure at $CD = h's$;

$$\therefore\ hs + x + y = h's,$$

$$\therefore\ x = (h'-h)s - y.$$

Now, by Marriotte's law, if a be the measure of the height of the bell,

$$\frac{y}{a} = \frac{h}{h'}, \quad \text{or,}\ y = \frac{h}{h'}a\ ;$$

$$\therefore\ x = (h'-h)s - \frac{h'}{h}a.$$

2.   *What must be the least size in cubic feet of an inflated balloon, that it may rise from the earth when filled with gas whose specific gravity compared with that of air is ·08, the weight of a cubic foot of air being ·3 grains, and the collapsed balloon car and contents weighing altogether 550 lbs.?*

Taking 1 as the measure of the specific gravity of air,

and V .... of the volume of the inflated balloon,

weight of inflated balloon, neglecting weight of envelope, $\Big\} = (·08 \times V \times ·3)$ grs.

weight of air displaced $= (·V \times 1)$ grs. $= V$ grs.

Now 1 cubic ft. of air weighs ·3 grs.,

$\therefore V$ .... ·3 V grs.;

$\therefore$ ascensional force $= (·3 V - ·08 V \times ·3)$ grs.

$$= (·92 \times ·3 V) \text{ grs.}$$

$\therefore$ ·92 × ·3 V = 550 × 7000,

$\therefore V = \dfrac{550 \times 7000}{·92 \times ·3}$ cub. ft. = 8072·5 cub. ft. nearly.

3.   *The weight of a globe in air is W, and in water w; find its radius, supposing s and a to be the specific gravities of water and air.*

Let R = radius of globe, and P = weight of globe in vacuo.

Then volume of globe $= \dfrac{4}{3} \pi R^3$;

$\therefore P - \dfrac{4}{3} \pi R^3 a = W$ .... (1),

$P - \dfrac{4}{3} \pi R^3 s = w$ .... (2).

Hence, subtracting (2) from (1),

$$\dfrac{4}{3} \pi R^3 (s - a) = W - w;$$

$\therefore R = \sqrt[3]{\left\{ \dfrac{3}{4\pi} \cdot \dfrac{W - w}{s - a} \right\}}.$

4. *How deep must a cylindrical diving bell be submerged so as to be just half full of water ?*

At first the bell is full of air of ordinary density.

When the bell is half full of water, the air is compressed into half its original volume, and therefore the density is doubled.

But the additional density is entirely due to the weight of a column of water 33 feet high.

Hence when the surface of the water in the bell is 33 feet below the upper surface, the bell will be half full of water.

5. *A spherical balloon is to be formed of a material of which the thickness is κ, and specific gravity relatively to air δ: if it be filled with gas of specific gravity d, prove that in order that it may ascend the extreme radius must exceed*

$$\frac{\kappa}{1-\left(\dfrac{\delta-1}{\delta-d}\right)^{\frac{1}{3}}}.$$

Let $x=$ extreme radius.

Then $x-\kappa=$ interior radius.

$\therefore$ weight of envelope alone $=\dfrac{4}{3}\pi\left\{x^3-(x-\kappa)^3\right\}\delta$ ... (1),

.... gas .... $\dfrac{4}{3}\pi(x-\kappa)^3 d$ .... (2),

.... air displaced ... $=\dfrac{4}{3}\pi x^3\times 1$ .... (3).

The balloon will not ascend unless the sum of (1) and (2) be less than (3).

$\therefore\dfrac{4}{3}\pi\left\{x^3-(x-\kappa)^3\right\}\delta+\dfrac{4}{3}\pi(x-\kappa)^3 d-\dfrac{4}{3}\pi x^3$ less than 0 ;

$\therefore x^3(\delta-1)$ less than $(x-\kappa)^3(\delta-d)$,

$\therefore\dfrac{x-\kappa}{x}$ greater than $\left(\dfrac{\delta-1}{\delta-d}\right)^{\frac{1}{3}}$,

$\therefore 1-\dfrac{\kappa}{x}$ greater than $\left(\dfrac{\delta-1}{\delta-d}\right)^{\frac{1}{3}}$,

$$\therefore \ \frac{\kappa}{x} \text{ less than } 1 - \left(\frac{\delta-1}{\delta-d}\right)^{\frac{1}{3}},$$

$$\therefore \ x \text{ greater than } \frac{\kappa}{1 - \left(\frac{\delta-1}{\delta-d}\right)^{\frac{1}{3}}}.$$

6. *For two given temperatures the readings of one thermometer are n° and m° and of another ν° and μ° respectively. What will be the reading of the latter when the former gives l° ?*

$(n-m)$ deg. of the 1st are equivalent to $(\nu-\mu)$ deg. of the 2nd.

$$\therefore \quad 1^0 \ \ldots\ldots\ldots\ldots \text{ 1st } \ldots\ldots\ldots\ldots\ldots\ldots \frac{\nu-\mu}{n-m} \ \ldots\ldots\ldots\ldots \text{ 2nd.}$$

$$\therefore \quad l^0 \ \ldots\ldots\ldots\ldots \text{ 1st } \ldots\ldots\ldots\ldots \left(\frac{\nu-\mu}{n-m}\right) l \ \ldots\ldots\ldots \text{ 2nd.}$$

7. *A globe, 2 feet in diameter, when floating is half immersed in water ; what is its weight ?*

The globe must be half as heavy as water.

Now volume of globe $= \frac{4}{3}\pi$ cubic feet,

and 1 cub. ft. of water weighs 62·5 lbs.

$$\therefore \ \frac{4}{3}\pi \text{ cub. ft. of water weigh } \left(62{\cdot}25 \times \frac{4\pi}{3}\right) \text{ lbs. ;}$$

$$\therefore \text{ weight of globe} = \frac{1}{2}\left(62{\cdot}25 \times \frac{4\pi}{3}\right) \text{ lbs.}$$
$$= 130{\cdot}9 \text{ lbs. nearly.}$$

8. *A sphere whose radius is 6 inches and weight 35 lbs. is suspended by a string. Required the tension of the string when the sphere is wholly immersed in water.*

$$\text{Volume of sphere} = \frac{4}{3}\pi\left(\frac{1}{2}\right)^3 \text{ cub. ft.} = \frac{\pi}{6} \text{ cub. ft.}$$

$$\text{Weight of water displaced} = \left(\frac{\pi}{6} \times 62{\cdot}5\right) \text{ lbs.}$$

$$\therefore \text{ tension of string} = \left(35 - \frac{\pi}{6} \times 62{\cdot}5\right) \text{ lbs.}$$
$$= 2{\cdot}275 \text{ lbs. nearly.}$$

9. *A pipe* 15 *feet long, closed at the upper extremity, is placed vertically in a tank of the same height, and the tank is filled with water. Prove that if the height of the water barometer be* 33 *ft.* 9 *in., the water will rise* 3 *ft.* 9 *in. in the tube.*

Let x = measure of height to which the water rises in feet.

Then $15 - x$ = measure of space filled with air.

By Marriotte's law, the pressure of the air inside may be represented by

$$\frac{15}{15-x} \times 33\frac{3}{4}.$$

But this pressure is also represented by the measure of a column of water $33\frac{3}{4}$ ft. + a column $(15-x)$ ft.

$$\therefore 33\frac{3}{4} + 15 - x = \frac{15}{15-x} \times 33\frac{3}{4},$$

or

$$x^2 - \frac{255}{4}x + \left(\frac{255}{8}\right)^2 = \frac{50625}{64},$$

$$\therefore x - \frac{255}{8} = \pm\frac{225}{8};$$

$$\therefore x = 60 \text{ ft. or } 3\frac{3}{4}\text{ ft.}$$

The first result is evidently impossible.

10. *If a lighter fluid rest upon a heavier, and their specific gravities be s and s', and if a body whose sp. gr. is σ rest with V of its volume in the upper fluid and V' in the lower, shew that*

$$V : V' = s' - \sigma : \sigma - s,$$

weight of body = weight of fluid displaced,

$$= \sigma(V + V'),$$
$$= sV + s'V',$$

$$\therefore V(\sigma - s) = V'(s' - \sigma),$$

$$\therefore V : V' = s' - \sigma : \sigma - s.$$

EXAMPLES. —IX.

1. Equal volumes of gold (s. g. 19·4) and silver (s. g. 10·4) balance on a straight lever, (1) in vacuo, (2) in water, (3) in mercury (s. g. 13·5). Find the ratio of the arms and position of the fulcrum in each case.

2. An inclined plane is immersed in a fluid (s. g. 3) and a body (s. g. 7) weighing 7 lbs. in vacuo is supported on the plane by a horizontal force of 3 lbs. Find the ratio of the height and base of the plane.

3. A balloon filled with Hydrogen (s. g. ·07) just rises in air (s. g. 1). The balloon, exclusive of the Hydrogen, weighs 10 cwt. If a cubic foot of air weigh 1·3 oz., find the volume of Hydrogen in the balloon, neglecting the volume of all else.

4. If the balloon in Question (3) rise and rest with its barometer at three-fourths of its original height, how much gas must have been expelled, and how much ballast thrown out?

5. Explain why the gas and ballast in Question (4) are expelled.

6. A cylindrical vessel is made of wood: the exterior radius is 4 inches and the interior 3 inches, the thickness of the bottom one inch, and the height of the cylinder 9 inches. It floats in water when the bottom is 3 inches below the surface. Find the specific gravity of the wood and the depth to which it will sink when a *small* hole is made in the bottom.

7. A piece of ice, supporting a stone, floats in a vessel of water. Will any change take place in the level of the water as the ice melts?

8. Shew that in a cylinder immersed as in Question (25) page 64, the depth of the interior surface below the exterior is a mean proportional between the height of the water in the cylinder and that of the water barometer.

9. A cubical water-tight box, whose edge is 1 foot, is sunk to a depth of 80 fathoms in the sea. Find the pressure on the top.

Would it make any difference in the circumstances of the box if it were not water-tight?

10. An elastic air-tight bag has forced into it air sufficient to fill 19 bags of the same original size. To what depth must it be sunk in the water that it may return to its original size, the height of the water-barometer being 34 feet?

11. A vessel made of thin heavy material and containing a cubic foot of fluid, the specific gravity of which is $\frac{7}{8}$, floats in water, the surfaces of the water and the fluid being in the same horizontal plane. Find the weight of the vessel when empty.

12. In Question (11) if some more fluid of the same kind be poured into the vessel, will the surface of the fluid or that of the water be the higher?

13. A cylinder 30 inches long is composed of lignum vitæ in its lower half and cork in its upper half, and floats vertically in water. If the specific gravities of lignum vitæ and cork be 1·1 and ·25 respectively, shew that the cylinder will float 20·25 inches deep.

14. Two pieces of cork, both small but the volume of one three times that of the other, are connected by a thread three feet long passing round a fixed pulley at the bottom of a tank of water 2 feet deep. Supposing the specific gravity of cork to be ·25, shew that in the position of equilibrium the smaller piece will be totally immersed and the larger piece half immersed.

15. Two reservoirs of water at different levels are separated by a solid embankment, and a bent iron tube of adequate length is placed with an end in each. If the barrel of an air-pump be screwed into an aperture at the top of the tube, shew that generally after sufficiently working the air-pump the water will flow through the tube from the higher reservoir to the lower. Under what circumstances will this fail to take place?

16. Two bodies of equal volume are placed one in each scale-pan of a Hydrostatic Balance, and are then immersed in two liquids which are such that the bodies just balance each other; the liquids are then interchanged, and it is found that the bodies balance when one of them is just half immersed. Find how much of the heavier body must be immersed in a liquid, composed of equal volumes of the two liquids, so that it may just balance the lighter not immersed.

17. A siphon ABC, each branch of which is less than 30 inches long, is filled with mercury and both ends are stopped. It is then placed with the end A in a bowl of mercury and the end C in a bowl of water, the surface of the mercury being *lower* than that of the water and higher than the end C. If the ends be simultaneously unstopped, shew that mercury will flow through the tube into the water provided that

$$\frac{z}{z'} \text{ be greater than } \frac{\rho'}{\rho},$$

z, z' being the respective depths of the end C below the planes of the surfaces, and ρ, ρ' the respective densities of mercury and water.

18. The air-vessel of a force-pump is a cylinder of height c, whose section A is the same as that of the piston: the water has to be lifted to height h of the water-barometer above the bottom of the air-vessel, by means of a pipe of section a and height h: if, when the pump commences working, the water be just below the valve in the air-chamber, find after how many strokes, each of length l, of the piston, the water will be at the top of the pipe.

19. A cylinder whose height is 8 inches, is floating with its axis vertical and its base 5 inches below the surface of water : a weight of 6 lbs. when placed on the top of the cylinder just brings the upper surface to the level of the water. Find the weight of the cylinder.

20. When two metals are mixed in equal volumes they form a compound of specific gravity 9 ; when they are mixed in equal weights they form a compound of specific gravity $8\frac{8}{9}$; find the specific gravities of the metals.

21. A cylindrical jar can just sustain a pressure of 165 lbs. to the square inch without breaking, and an air-tight piston which fits the jar is thrust down and compresses the air in the jar. Find the height of the jar, supposing it to burst when the piston is an inch from the bottom of the cylinder, the pressure of atmospheric air being 15 lbs. to the square inch.

22. In Smeaton's air-pump if there be communication with a condenser through the upper valve, and the capacity of the cylinder be half that of either receiver, compare the pressures in the receivers after two descents and ascents of the piston.

Notes.

1. On page 6, line 7, insert the word *additional* before the word *force*, so that it reads thus: "the same additional force *P*."

2. On page 15 the construction of the cylinder and lines 6, 7, 8, 9 are not *necessary* to the proof, for it follows at once from Art. 34 that

fluid pressure at *A* = fluid pressure at *B*.

3. On page 24 it might be clearer if we inserted the sign × or the word *times* between *VS*, and (unit of weight) in line 7, also between $\frac{W}{V}$ and (unit of specific gravity) in line 14, and so in several other cases in pages 24 and 25.

4. The first sentence in page 53 is not quite correct: it might better stand thus: "The exhaustion of the air is retarded by the difficulty of making the piston come into close contact with the valves at *A* and *C*, and it must always be limited by the weight of the valve *C*."

5 The Aneroid Barometer is so called because *no liquid* (*à* privative and *νηρός* "moist") is used in its construction. A metal cylinder about an inch in height, closed by an elastic piece of metal, is exhausted, and as the metal covering rises or is depressed, according to the changes of atmospheric pressure, it sets in motion hands like those of a watch connected with it.

6. In reading the descriptions of the Pumps in pages 67—71 the student must be careful not to derive any erroneous notions from the use of the words *Suction*-pipe. It is retained (perhaps not wisely) as a technical term, convenient for distinguishing the lower part of the pumps from the barrel.

7. In the description of the Siphon on page 72 it is said to be of *uniform* bore. This is not essential to the working of the instrument, but it conduces to the regular action of it, and renders the explanation more simple.

It is also stated on page 72 that the *longer branch must be outside* the vessel. This is not necessary, for the instrument will work with the shorter branch outside, provided that the extremity of that branch be below the surface of the fluid.

8. To the Thermometers it might be well to add that which is called De Lisle's. This is much used in Russian scientific operations. In it the boiling point is marked 0°, and the freezing point 150°.

9. It should be carefully observed that the freezing point of a Thermometer is found by placing the instrument not in *freezing water*, but in *melting ice*.

ANSWERS.

Examples I. (page 8.)

1. $56\frac{6}{7}$ tons. 2. 30 tons. 3. $29629\cdot\dot{6}2\dot{9}$ lbs.
4. 1 oz. 5. 1 oz.
6. The area of a circle whose radius is r is πr^2, and taking $\frac{22}{7}$ as an approximate value of π, the answer is $5587\frac{13}{49}$ cwt.

Examples II. (page 18.)

1. 20 lbs. 2. $37\frac{7}{12}$ lbs. 3. $7:6$. 4. $9:8$.
5. 10 feet. 6. 12 lbs. 7. 9 lbs.
8. 1 ton 7 cwt. 3 qrs. 17 lbs. 9. 11 lbs. $12\frac{8}{11}$ oz.
10. 22500 lbs. 11. $1125\sqrt{3}$ lbs. 12. $\frac{2}{3}$ of its height.

13. Since the external pressure on the cork *increases* with the depth, while the internal pressure is *constant*, the cork will be forced in when the former exceeds the latter.

14. $12\frac{1}{2}$ tons. 15. 18 feet.

Examples III. (page 27.)

1. 165 lbs. 2. $18:1$. 3. $7\frac{13}{32}$ oz. 4. $\cdot8$.
5. 5 oz. 6. $1\frac{43}{60}$ oz. 7. $\dfrac{\cdot016n}{m}$. 8. $7\cdot776$.
9. $1\cdot1\dot{6}$. 10. $\cdot844$. 11. 14.
12. $\dfrac{3}{4}$ cub. in.; $\dfrac{5}{4}$ cub. in.
13. Volumes as $1:2$, weights as $1:4$. 14. $2:1$.
15. $2\frac{17}{18}$. 16. 2. 17. $9\cdot325$.

18. If d_1, d_2, d_3 be the measures of the densities of the fluids, and d be the measure of the density of the mixture, $d_3 = 3d - d_1 - d_2$.

19. 6·241...   20. ·802... 21. 18·41.   22. 1·61...

23. 3·13̇.   24. 8·6... oz.

25. The volumes are as 57 : 1, the weights as 2223 : 97.

Examples IV. (page 42.)

1. 3·3̇.   2. 507870 tons.   3. three-fourths.

4. 4 dwts. 20$\frac{18}{37}$ grs. 5. 4. 6. 3 times weight of tub.

7. two-thirds.   8. $\frac{52}{57}$ oz. 9. $\frac{65}{432}$ oz. 10. 3 oz.

11. 42 oz. 12. $\frac{5}{36}$ oz. 13. $\frac{5}{6}$ lbs. 14. 3·5.

15. $\frac{60}{73}$. 16. $\frac{6}{7}$. 17. $\frac{75}{91}$. 18. 17$\frac{1}{2}$. 19. 2$\frac{1}{7}$.

20. $\frac{16}{19}$. 21. $\frac{1362}{2731}$. 22. 2. 23. 47$\frac{1}{2}$ lbs.

24. 12 feet.   25. ·66.

26. Because the specific gravity of salt water is greater than that of fresh water.

27. $\frac{34}{63}$ inches.   28. 1728.   29. 7 lbs. 9$\frac{1}{13}$ oz.

30. Edge of cube is 2 feet.   31. 5$\frac{2}{7}\frac{1}{1}$.

32. $\dfrac{\text{height of triangle}}{\sqrt{3}}$. 33. 1 : 4 when vertex is downwards ; 3 : 4 when vertex is upwards. 34. 16 lbs.

35. ·75. 36. 2 inches. 37. ·9. 38. $\frac{26}{135}$ inch.

39. 4 lbs.   40. 3 lbs.   41. 95 lbs.

42. $w_1(w_2 - w) : w_2(w_1 - w)$. 43. Increased, if the wood be lighter than water. 44. 3 : 2. 45. 1·1.

47. $\frac{14129}{17740}$ or ·8 nearly.   48. 5$\frac{5}{13}$ inches.

49. 8$\frac{19}{81}$ lbs. 50. 9$\frac{14}{30}$ lbs. 51. 10$\frac{10}{13}$. 52. 1·3̇.

53. 6 inches.   55. 2$\frac{2}{8}$.

56. $\frac{86400}{2053}$ cub. in.   57. $\frac{1}{666}$ of a cubic yard.

58. $\frac{3}{5}$ of volume.　　59. 750 oz.　　60. $\frac{85}{103}$ of a cub. ft.

61. 936302451·687 cub. ft.　　62. $\frac{W}{3}$.　　63. $m:n$.

64. 900 grains.　　65. $1\frac{13}{247}$ or 1·0272 nearly.　　66. ·54.

67. 433 grains.　　　　　　69. $42\frac{174}{1053}$ cub. in.

Examples V. (page 60.)

1. Density $=\left(\dfrac{10}{11}\right)^{6}$ times original density.　2. $\dfrac{64}{81}$ times original pressure.　　3. 9 : 1.　　4. No: because the pressure varies with the depth alone; so that if the section varied there would still be equal vertical increments of space for equal increments of pressure.　　5. 53¾ lbs.
6. $1\frac{13}{20}$ inches.　　7. The mercury would fall to the level of the surface in the cup.　　8. 14·625 lbs.　　9. 10 feet.
10. No: because a volume of mercury equal to that displaced by the iron will descend and allow the iron to take its place without disturbing the general upper surface.
11. Sink: see answer to (16).　　12. The mercury would descend a little.　　13. 2·38 square inches.　　14. ·0109 of original volume.　　15. 1 ft. $5\frac{931}{1367}$ in.　　16. When the floating body is partially immersed, both air and water are displaced: but the *absolute* weight of floating body = weight of displaced fluids, which must therefore be constant: therefore when the barometer rises, there must be a less water displacement, i.e. the body rises: while any decrease in the atmospheric pressure (when the barometer falls) will necessitate an increased water displacement, and therefore the body then sinks a little.　　17. 1 : 2.　　20. 28·8 inches.
21. 1080 lbs.　　22. 26$\frac{11}{17}$ inches.　　23. 5 feet.
24. 6 : 1.　　25. The air will be compressed inside, and so displace less water: and since it floated originally, it will now sink, because the weight of displaced fluid is now less than the weight of the body.　　26. 5 times original pressure.
27. 32 ft.　　28. $5\frac{85}{108}$ oz.　　29. 4776 oz.
30. $AB:BC=p:q-p$.　　32. The space between zero

point and any graduation ought to be less than the space indicated by the number placed against that graduation in the ratio of 17 : 18.   33. 18·1505 lbs. nearly.

34. $\frac{9}{10}$ of an inch.   35. Low.   36. $4\frac{2}{3}$ inches.

Examples VI. (page 75.)

1. It will increase the time of filling the receiver, since the only *effective* work would be done by the descending piston, after passing the hole. It will fill the tank in 3 times the original time.   2. $27\frac{7}{8}$ lbs.

3. (a) If the hole be below the level of short end, no effect.

(β) If above this level but still in the long branch, all the fluid in this branch below the hole will descend, and all above in the same branch will ascend, causing the remainder of the fluid to flow through the short branch, till the siphon is emptied.

(γ) If in the short branch, all the fluid below the hole is this branch will descend ; all above in the same branch will ascend and flow through the long branch, emptying the siphon.

(δ) If at the top of the siphon, the fluid will descend in each branch and empty the siphon.

4. 32 ft. 9·53 in. or 32·79416 feet.   5. The fluid would descend in each branch and the siphon be emptied.

6. Equally well at both, if the siphon be not too high.

7. No: because the hold is *lower* than the surface in harbour.

8. 33 ft. 11·1 in.   9. If the air be removed from the siphon, the fluids would first ascend in each branch and afterwards flow as usual.   10. The water would rise in the inverted tube as high as the top of the inserted tube and afterwards flow out of it.   11. First, the water would soon cease to flow. Secondly, it would rise in each branch, and afterwards flow.   12. (a) The water will flow into the lower vessel. (β) The water will descend in each branch till it stands at 34 feet above each surface. (γ) The same as (a).
13. Each branch 2 feet.

Examples VII. (page 81.)

1. (1) $-1\frac{1}{9}^{0}$; $-\frac{8}{9}^{0}$.   (2) $7\frac{2}{9}^{0}$; $5\frac{7}{9}^{0}$.   (3) $13\frac{1}{9}^{0}$; $10\frac{2}{9}^{0}$.
(4) $-17\frac{7}{9}^{0}$; $-14\frac{2}{9}^{0}$.   (5) $-21\frac{2}{9}^{0}$; $-17\frac{1}{9}^{0}$.   (6) $-42\frac{7}{9}^{0}$; $-34\frac{2}{9}^{0}$.
2. (1) $6\frac{1}{4}^{0}$; $43\frac{1}{4}^{0}$. (2) 25^{0}; 77^{0}. (3) 0^{0}; 32^{0}. (4) $-22\frac{1}{2}^{0}$; $-8\frac{1}{2}^{0}$.
(5) -80^{0}; -112^{0}.   (6) 150^{0}; 302^{0}.   3. (1) $60\frac{1}{8}^{0}$; $12\frac{1}{8}^{0}$.
(2) 113^{0}; 36^{0}.   (3) 230^{0}; 88^{0}.   (4) 32^{0}; 0°.   (5) 5^{0}; -12^{0}.
(6) $-11\frac{1}{8}^{0}$; $-19\frac{1}{8}^{0}$.   4. Yes: if the graduations are to be uniform.   5. 10^{0} Cent. and 50^{0} Fah.   6. 10^{0} Cent. and 50^{0} Fah.   7. -40^{0}.   8. Make each degree $\frac{4}{5}$ ths that on Fahrenheit.   8. $25\frac{5}{8}^{0}$.   10. 20^{0} Cent, 68^{0} Fah.   11. $9\frac{1}{9}^{0}$. 12. The graduations would be inconveniently small.
13. 80^{0} Fah.   14. 20^{0} Cent., 68^{0} Fah.   15. $-11\frac{3}{4}$ Cent., $11\frac{3}{4}$ Fah.   16. 24^{0}.   17. 23^{0}.   18. 59^{0} Fah., 12^{0} Reaum.; if d be the number of degrees, Fah. rises $\frac{9d}{5}$ and Reaum. $\frac{4d}{5}$.

Examples VIII. (page 87.)

1. $4 : 37$.   2. $\cdot495$.   3. -40^{0} and 662^{0}.
4. $11\cdot8125$ inches. 5. $1 : p$.   6. $1\cdot08\dot{3}$.
7. 31 ft. $8\cdot8$ in.   8. $-2\frac{7}{9}^{0}$.   9. $22\frac{1}{2}$ ft.
10. $\frac{1}{80}$ cub. ft.   11. $137\frac{1}{2}^{0}$.   12. $\frac{2400}{13}$.

13. 10 feet.   14. (a) $2\frac{1}{2}$ lbs. downwards;

(β) $2\frac{6}{7}$ lbs. upwards.   15. $7\frac{1}{2}$ ft., and $13\frac{3}{4}$ ft.
16. 50^{0} Fahrenheit.   17. 122^{0} Fahrenheit.
18. $15\frac{9}{17}$.   19. 2.   20. $17\frac{3}{4}$ inches from one end.

Examples IX. (page 94.)

1. (1) $97 : 52$.   (2) $92 : 47$.   (3) $59 : 90$.
In (3) fulcrum is at one end, and gold between fulcrum and silver.

2. 3 : 4.   3. $\dfrac{10 \times 112 \times 16}{1 \cdot 20}$ ft.   4. $\dfrac{1}{3}$ of the gas

has been expelled, and $\dfrac{1}{4}$ of the whole weight thrown out.

5. *Gas* to preserve equilibrium of *internal* and *external* pressures on the balloon. *Ballast* to preserve equilbrium of *vertical* pressures on the balloon.

6. Sp. gr. $= \dfrac{2}{3}$. Height immersed $= 5\dfrac{4}{7}$ inches.

7. No change will take place till the stone falls from the ice, it will then displace less water than before, and the surface will consequently sink.

9. Taking a cubic foot of water to weigh 1000 oz., the resultant pressure is 30000 lbs. The pressure would be the same inside as outside.

10. 102 fathoms.   11. 125 oz.   12. The fluid surface.

16. $\dfrac{2}{7}$.   18. $\dfrac{A\,(2h + c - \sqrt{c^3 + 4h^2}) + a\,(\sqrt{c^2 + 4h^2} - c)}{2Ac}$.

19. 10 lbs.   20. 10 and 8.   21. 11 inches.

22. 33 : 8.